High-Performance Computing on the Intel® Xeon Phi™

Endong Wang • Qing Zhang • Bo Shen
Guangyong Zhang • Xiaowei Lu
Qing Wu • Yajuan Wang

High-Performance Computing on the Intel® Xeon Phi™

How to Fully Exploit MIC Architectures

 Springer

Endong Wang
Qing Zhang
Bo Shen
Guangyong Zhang
Xiaowei Lu
Qing Wu
Yajuan Wang
Inspur, Beijing, China

Translators
Dave Yuen
Chuck Li
Larry Zheng
Sergei Zhang
Caroline Qian

www.waterpub.com.cn

Copyright © 2012 by China Water & Power Press, Beijing, China
Title of the Chinese original: MIC 高性能计算编程指南
ISBN: 978-7-5170-0338-0
All rights reserved

ISBN 978-3-319-35879-6   ISBN 978-3-319-06486-4 (eBook)
DOI 10.1007/978-3-319-06486-4
Springer Cham Heidelberg New York Dordrecht London

© Springer International Publishing Switzerland 2014
Softcover reprint of the hardcover 1st edition 2014

Printed on acid-free paper

Springer is part of Springer Science+Business Media (www.springer.com)

Foreword by Dr. Rajeeb Hazra

Today high-performance computing (HPC), especially the latest massively supercomputers, has developed quickly in computing capacity and capability. These developments are due to several innovations. Firstly, Moore's law, named after the Intel founder Gordon Moore, predicts that the number of semiconductor transistors will double every 18–24 months. According to Moore's law, Intel continues to improve performance and shrink the size of transistors as well as to reduce power consumption, all at the same time. Another innovation is a series of CPU-improving microstructures which ensure that the performance of a single thread coincides with the parallelism in each successive CPU generation.

The development of HPC plays an important role in society. Although people are inclined to pay more attention to great scientific achievements such as the search for the Higg's boson or the cosmological model of cosmic expansion, the computing capability that everyone can now acquire is also impressive. A modern two-socket workstation based on the Intel® Xeon series processors could show the same performance as the top supercomputer from 15 years ago. In 1997, the fastest supercomputer in the world was ASCI Red, which was the first computing system to achieve over 1.0 teraflops. It had 9298 Intel Pentium Pro processors, and it cost $55,000,000 per teraflop. In 2011, the cost per teraflop was reduced to less than $1000. This reduction in cost makes high-performance computing accessible to a much larger group of researchers.

To sufficiently make use of the ever-improving CPU performance, the application itself must take advantage of the parallelism of today's microprocessors. Maximizing the application performance includes much more than simply tuning the code. Current parallel applications make use of many complicated nesting functions, from the message communication among processors to the parameters in threads. With Intel CPUs, in many instances we could achieve gains of more than ten times performance by exploiting the CPU's parallelism.

The new Intel® Xeon Phi™ coprocessor is built on the parallelism programming principle of the Intel® Xeon processor. It integrates many low-power consumption cores, and every core contains a 512-bit SIMD processing unit and many new vector instructions. This new CPU is also optimized for performance per watt. Due to a computing capability of over one billion times per second, the Intel® Xeon Phi™ delivers a supercomputer on a chip. This brand new microstructure delivers ground-breaking performance value per watt, but the delivered performance also relies on those applications being sufficiently parallelized and expanded to utilize many cores, threads, and vectors. Intel took a new measure to release this parallelism ability. Intel followed the common programming languages (including C, C++, and Fortran) and the current criteria. When readers and developers learn how to optimize and make use of these languages, they are not forced to adopt nonstandard or hardware-dependent programming modes. Furthermore, the method, based on the criteria, ensures the most code reuse, and could create the most rewards by compiling the transplantable, standardized language and applying it to present and future compatible parallel code.

In 2011, Intel developed a parallel computing lab with Inspur in Beijing. This new lab supplied the prior use and development environment of the Intel® Xeon processor and Intel® Xeon Phi™ coprocessor to Inspur Group and to some excellent application developers. Much programming experience can be found in this book. We hope to help developers to produce more scientific discovery and creation, and help the world to find more clean energy, more accurate weather forecasts, cures for diseases, develop more secure currency systems, or help corporations to market their products effectively.

We hope you enjoy and learn from this venerable book. This is the first book ever published on how to use the Intel® Xeon Phi™ coprocessor.

Santa Clara, CA   Rajeeb Hazra

Foreword by Prof. Dr. Rainer Spurzem

Textbooks and teaching material for my Chinese students are often written in English, and sometimes we try to find or produce a Chinese translation. In the case of this textbook on high-performance computing with the Intel MIC, we now have a remarkable opposite example—the Chinese copy appeared first, by Chinese authors from Inspur Inc. led on by Wang Endong, and only some time later can all of us English speakers enjoy and benefit from its valuable contents. This occasion happens not by chance—two times in the past several years a Chinese supercomputer has been the certified fastest supercomputer in the world by the official Top500 list (http://www.top500.org). Both times Chinese computational scientists have found a special, innovative way to get to the top—once with NVIDIA's GPU accelerators (Tianhe-1A in 2010) and now, currently, with the Tianhe-2, which realizes its computing power through an enormous number of Intel Xeon Phi hardware, which is the topic of this book. China is rapidly ascending on the platform of supercomputing usage and technology at a much faster pace than the rest of the world. The new Intel Xeon Phi hardware, using the Intel MIC architecture, has its first massive installation in China, and it has the potential for yet another supercomputing revolution in the near future. The first revolution, in my opinion, has been the transition from traditional mainframe supercomputers to Beowulf PC clusters, and the second was the acceleration and parallelization of computations by general-purpose computing on graphical processing units (GPGPU). Now the stage is open for—possibly—another revolution by the advent of Intel MIC architecture. The past revolutions of accelerators comprised a huge qualitative step toward better price–performance ratio and better use of energy per floating point operation. In some ways they democratized supercomputing by making it possible for small teams or institutes to assemble supercomputers from off-the-shelf components, and later even (GPGPU) provide massively parallel computing in just a single desktop. The impact of Intel Xeon Phi and Intel MIC on the market and on scientific supercomputing has yet to be seen. However, already a few things can be anticipated; and let me add that I write this from the perspective of a current heavy user and provider of GPGPU capacity and capability. GPGPU architecture, while it provides outstanding performance for a fair range of applications, is still not as common as expected a few years ago. Intel MIC, if it fulfills the promise of top-class performance together with compatibility to a couple of standard programming paradigms (such as OpenMP as it works on standard Intel

CPUs, or MPI as it works on standard parallel computers) may quickly find a much larger user community than GPU. I hope very much that this very fine book can help students, staff, and faculty all over the world in achieving better results when implementing and accelerating their tasks on this interesting new piece of hardware, which will for sure appear on desktops, in institutional facilities, as well as in numerous future supercomputers.

Beijing, China   Rainer Spurzem

Foreword by Endong Wang

Currently scientists and engineers everywhere are relentlessly seeking more computing power. The capability of high-performance computing has become the competition among the few powerful countries in the world. After the "million millions flops" competition ended, the "trillion flops" contests have begun. The technology of semiconductors restricts the frequency of processors, but multi-processors and the many-integrated processors have become more and more important. When various kinds of many-integrated cores came out, we found that although the high point of computing has increased a lot, the compatibility of the applications became worse, and the development of applications has become more complicated. A lack of useful applications would render the supercomputer useless.

At the end of 2012, Intel corporation brought out the Intel® Xeon Phi™ coprocessor based on the many-integrated core. This product integrated more than 50 cores that were based on the x86 architecture into one PCI -Express interface card. It is a powerful supplement to the Intel® Xeon CPU, and brings a new performance experience for a highly parallelized workload. It is easy to program on this product, and there's almost no difference when compared with traditional programming. The code on the Intel® Xeon Phi™ coprocessor could be applied to a traditional platform based on CPU without any modifications, which protects the user's software investment. It can supply hundreds of running hardware threads, which could bring high parallelism and meet the current demands of high parallelization.

The Inspur-Intel China Parallel Computing Joint Lab was found on August 24, 2011. This lab aims to promote the trillion-flops supercomputing system architecture and application innovation, establish the ecological condition of high-performance computing, and accelerate supercomputing in China into the trillion-flops era. The research and innovation in the Inspur-Intel China Parallel Computing Joint Lab will make a positive impact on the development of supercomputing in China in the next ten years, especially in the beginning of the trillion-flops era for the rest of the world. The Inspur-Intel China Parallel Computing Joint Lab contributed to the completion of the Intel® Xeon Phi™ coprocessor and made a tremendous effort to popularize it.

This book was finished by several dedicated members of the Inspur-Intel China Parallel Computing Joint Lab. In this book, relevant knowledge about the Intel® Xeon Phi™ coprocessor, programming methods in using the Intel® Xeon Phi™

coprocessor, optimizations for the program, and two successful cases of applying the Intel® Xeon Phi™ coprocessor in practical high-performance computing are introduced. This book has a clear structure and is easy to understand. It contains a programming basis, optimization, and specific development projects. At the same time, a lot of figures, diagrams, and segments of program were included to help readers understand the material. The authors of this book have plenty of project experience and have added their practical summaries of these projects. So this book not only introduces the theory, but it also connects more closely to actual programming. This book is also the first to introduce the Intel® Xeon Phi™ coprocessor and embodies the achievement of these authors. We hope to see China accumulate some great experience in the field of HPC. The authors and the members of the Inspur-Intel China Parallel Computing Joint Lab made great efforts to ensure the book publishing coincides with the Intel® Xeon Phi™ coprocessor, and they should be respected for this.

We hope the readers will grasp the full use of the Intel® Xeon Phi™ coprocessor quickly after reading this book, and gain achievements in their own fields of HPC application by making use of the Intel® Xeon Phi™ coprocessor. The Inspur Group hopes to dedicate themselves to HPC endeavors together with Intel Corporation.

Beijing, China   Endong Wang

Preface

High-performance computing (HPC) is a recently developed technology in the field of computer science, and now computational science. HPC can secure a country's might, improve its national defense science, and promote the rapid development of highly sophisticated weapons. HPC is one of the most important measures of a country's overall prowess and economic strength. With the rapid growth of an information-based society, people are demanding more powerful capabilities in information processing. HPC is used not only for oil exploration, weather prediction, space technology, national defense, and scientific research, but also in finance, government, education, business, network games, and other fields that demand more computing capability. The drive in research to reach the goal of "trillion flops" computing has begun, and people are looking forward to solving larger scale and more complicated problems by using a trillion-flops supercomputer.

In this century, the many-integrated core (MIC) era has finally arrived. Today, the HPC industry is going through a revolution, and parallel computing will be the trend of the future as a prominent hot spot for scientific research. Current mainstream research has adopted the CPU-homogeneous architecture, in which there are dozens of cores in one node; this is not unusual. In large-scale computing, thousands of cores will be needed. Meanwhile, the CPU-homogeneous architecture faces a huge challenge because of its low performance-to-power ratio, performance-to-access memory ratio, and low parallel efficiency. When computing with the CPU+GPU heterogeneous architecture, the MIC acceleration technology of GPU is used. More and more developers have become dedicated to this field, but it also faces challenges such as fined-grained parallel algorithms, programming efficiency, and performance on a large scale. This book focuses on the central issues of how to improve the efficiency of large-scale computing, how to simultaneously shorten programming cycles and increase software productivity, and how to reduce power consumption.

Intel Corporation introduced the Intel® Xeon Phi™ series products, which are based on the MIC, to solve highly parallelized problems. The performance of the double-precision of this product has reached teraflop levels. It is based on the current x86 architecture, and supports OpenMP, pThread, MPI, and many parallel programming models. It also supports the traditional C/C++/Intel® Cilk™ Plus, Fortran, and many other programming languages. It is programmed easily, and many associated tools are supported. For applications that are difficult to realize by the traditional CPU platform, the MIC platform will greatly improve performance, and the source code can be shared by the CPU and MIC platform without any modifications. The combination of CPU and MIC in the x86 platform in heterogeneous computing provides HPC users with a new supercomputing solution.

Since the Inspur-Intel China Parallel Computing Joint Lab was established on August 24, 2011, the members have dedicated themselves to HPC application programs on the MIC platform, and have ensured that the Intel® Xeon Phi™ series products would be released smoothly. We have accumulated a large amount of experience in exploring the software and hardware of MIC. It is a great honor for us to participate in the technology revolution of HPC and introduce this book to readers as pioneers. We hope more readers will make use of MIC technology and enjoy the benefits brought forth by the Intel® Xeon Phi™ series products.

Target Audience

The basic aim of this book is to help developers learn how to efficiently use the Intel® Xeon Phi™ series products, by which they can develop, transplant, and optimize their parallel programs. The general content of this book introduces some computing grammar, programming technology, and optimization methods in using MIC, and we also offer some solutions to the problems encountered during actual use based on our optimization experience.

We assume that readers already have some basic skills in parallel programming, but have a scant knowledge of MIC. This book does not intend to introduce the theory of parallel computing or algorithms, so we also assume that readers already have this knowledge. In spite of this, when faced with the parallel algorithm, we still describe it in a simple way. We assume that readers are familiar with OpenMP, MPI, and other parallel models, but we also state the basic grammar. We assume that readers can make use of any one of the C/C++/Fortran languages, and that C/C++ is preferred. However, the ideas and advice stated in this book are also adapted to other high-level languages. Moreover, when the Intel® Xeon Phi™ series of products support other languages in the future, most of the optimization methods and application experience will still be effective. Generally speaking, this book is for three types of computing-oriented people:

Students and professional scientists and engineers in colleges, universities, and research institutes, and developers engaged in parallel computing, multi-core, and many integrated core technology.

IT employees, especially those who develop HPC software, improve application performance by many-integrated cores, and pursue extreme performance in the HPC field.

HPC users in other fields, including oil exploration, biological genetics, medical imaging, finance, aerospace, meteorology, and materials chemistry. We hope to help them to improve the original CPU performance by means of MIC and ultimately increase productivity.

We wish to benefit more readers with this book. In the future we also hope to engage more and more readers around the world.

About This Book

Because of the diverse characteristics of MIC architecture, this book cannot be sorted strictly into well-defined sections. This book introduces the MIC programming language and Intel® Xeon Phi™ series products, and it also describes optimization in parallel programming. Through this book, we hope readers will fully understand MIC, and we expect readers to make good use of MIC technology in future practice.

This book includes three parts. The first one covers MIC basics, and includes Chaps. 1–7, in which fundamental knowledge about the MIC technology is introduced.

In Chap. 1, the development of parallel computing is recalled briefly. The current hardware characteristics of parallel computing are compared. Then MIC technology is introduced, and the advantages of MIC are stated.

In Chap. 2, the hardware and software architecture of MIC are introduced. Although there's no influence on programming on MIC in the absence of this background knowledge, exploring the MIC architecture deeply will help our programs become more adapted to MIC.

In Chap. 3, by computing the circumference ratio pi, the characteristics of MIC programming are directly demonstrated to readers. In addition, we introduce the background procedures of the program.

In Chap. 4, the background knowledge of MIC programming is discussed, including the basic grammar of OpenMP and MPI. If you have had this basic training, you can skip this chapter altogether.

In Chap. 5, the programming model, grammar, environment variables, and compilation options of MIC are introduced. You should be able to grasp the method of writing your own MIC program by this chapter.

In Chap. 6, some debugging and optimization tools and their usage are introduced. These tools bring a great deal of convenience to debugging and optimization.

In Chap. 7, some Intel mathematical libraries that have been adapted on MIC are discussed, including VML, FFT, and Blas.

The second section covers performance optimization, and comprises Chaps. 8 and 9.

In Chap. 8, the basic principles and strategy of MIC optimization are introduced, and then the methods and circumstance of MIC optimization are stated. The general methods of MIC optimization are covered. Moreover, most of the methods are applicable to the CPU platform, with a few exceptions.

In Chap. 9, through the classical example in parallel computing—the optimization of matrix multiplication—the optimization measures are stated step-by-step in the method of integrating theory with practice.

The third and last section covers project development, and includes Chaps. 10 and 11.

In Chap. 10, we propose a set of methods to apply parallel computing to project applications by summarizing our experiences on development and optimization of our own projects. We also discuss how to determine if a serial or parallel CPU program is suitable for MIC, and how to transplant the program onto MIC.

In Chap. 11, we show, using two actual cases of how the MIC technology influences an actual project.

In the early stages, this book was initiated by Endong Wang, the director of the State Key Laboratory of high-efficiency server and storage technology at the Inspur-Intel China Parallel Computing Joint Lab, and the senior vice president of Inspur Group Co., Ltd. Qing Zhang, the lead engineer of the Inspur-Intel China Parallel Computing Joint Lab, formulated the plan, outline, structure, and content of every chapter. Then, in the middle stage, Qing Zhang organized and led the team for this book, checking and approving it regularly. He examined and verified the accuracy of the content, the depth of the technology stated, and the readability of this book, and gave feedback for revisions. This book was actually written by five engineers in the Inspur-Intel China Parallel Computing Joint Lab: Bo Shen, Guangyong Zhang, Xiaowei Lu, Qing Wu, and Yajuan Wang. The first chapter was written by Bo Shen. The second chapter was written by Qing Wu and Bo Shen. The third through fifth chapters were written by Bo Shen, and Yajuan Wang participated. The sixth chapter was written by Qing Wu. The seventh chapter was written by Xiaowei Lu. The eighth chapter was written by Guangyong Zhang, and Bo Shen and Yajuan Wang participated. The ninth chapter was written by Guangyong Zhang. The tenth chapter was written by Bo Shen. The eleventh chapter was written by Xiaowei Lu and Guangyong Zhang. In the later stage, this book was finally approved by Endong Wang, Qing Zhang, Dr. Warren from Intel, and Dr. Victor Lee.

The whole source code has been tested by the authors of this book, but because of the initial stage of MIC technology, we cannot ensure that these codes will be applicable in the latest release. Hence, if any updates come out for the compiler and the execution environment of MIC, please consult the corresponding version manual by Intel.

Acknowledgments

The publication of this book is the result of group cooperation. We would like to show our respect to the people who gave their full support to the composition and publication.

We must express our heartfelt thanks to Inspur Group and Intel Corporation, who gave us such a good platform and offered the working opportunity in the

Inspur-Intel China Parallel Computing Joint Lab. We are fortunate to be able to do research on MIC technology.

We are grateful for the support of the leadership of Inspur Group, especially to the director of the HPC Center, Inspur Group, Jun Liu, who supplied us with financial support and solicitude.

We are grateful to Michael Casscles, Dr. Wanqing He, Hongchang Guo, Dr. David Scott, Xiaoping Duan, and Dr. Victor Lee for their support of the technology and resources for our daily work in the parallel computing joint lab. We especially can't forget Wanqing! He supplied us with plenty of guidance from experience before writing this book. We are also grateful to Dr. Raj Hazra, GM Technical Computing in Intel Corporation, and Joe Curley, MD Technical Computing in Intel Corporation, for their support of the Inspur-Intel China Parallel Computing Joint Lab.

We are grateful to our application users: BGP Inc., China National Petroleum Corp, Institute of Biophysics, Chinese Academy of Sciences, Northwestern Polytechnical University, Chinese Academy of Meteorological Sciences, and Shandong University—especially Prof. Fei Sun and Dr. Kai Zhang from the Institute of Biophysics Chinese Academy of Sciences—and Profs. Chengwen Zhong and Qinjian Li from Northwestern Polytechnical University. The cases in this book come from them.

We are grateful to Inspur Group and Intel Corporation for their support, especially the managers Yongchang Jiang and Ying Zhang from the High-Efficiency Server Department, Inspur Group, who were able to save us a great deal of time.

We thank very much Dr. Haibo Xie and Xiaozhe Yang; we are unable to forget this pleasant time.

We are grateful to the families of the authors for their consideration and patience.

We thank the editors from China WaterPower Press, especially to the editor Chunyuan Zhou and editor Yan Li for their tolerance of our demands. This book could not possibly be published without their hard work.

We are very grateful for the English translation made by Professor David A. Yuen and his team from the University of Minnesota, Twin Cities, China University of Geosciences, Wuhan, consisting of Qiang (Chuck) Li, Liang (Larry Beng) Zheng, Siqi (Sergei) Zhang, and Caroline Qian. Jed Brown and Karli Rupp from Argonne National Laboratory also gave very useful advice, and finally, Prof. Xiaowen Chu and Dr. Kayiyong Zhao from Hong Kong Baptist University are to be thanked for their help in proofreading of the last few chapters.

Lastly, we are grateful to all the others whom we have not acknowledged.

MIC technology has just come out, so there are undoubtedly some mistakes to be found in this book. We apologize for this and look forward to any suggestions from our readers. This is the first book ever written in any language on MIC technology; it was published in the fall of 2012, and is to be contrasted with the newer books coming out from the USA in 2013 bearing the names of Intel Xeon Phi coprocessor.

Beijing, China   Qing Zhang

Contents

Introduction to the Authors

Endong Wang is both a Director and Professor of the Inspur-Intel China Parallel Computing Joint Lab in Beijing, China. He has received a special award from the China State Council, and is also a member of a national advanced computing technology group of 863 experts, the director of the State Key Laboratory for high-efficiency server and storage technology, Senior Vice President of the Inspur group, the chairman of the Chinese Committee of the International Federation for Information Processing (IFIP), and Vice President of the China Computer Industry Association. He is the winner of the National Scientific and Technology Progress Award as the first inventor in three projects, the winner of the Ho Leung Ho Lee Science and Technology innovation award in 2009, and has garnered 26 national invention patents.

Qing Zhang has a master's degree in computer science from Huazhong Technical University in Wuhan and is now a chief engineer of the Inspur-Intel China Parallel Computing Joint Lab. He is manager of HPC application technology in Inspur Group—which engages in HPC, parallel computing, CPU multi-core, GPU, and MIC technology—and is in charge of many heterogeneous parallel computing projects in life sciences, petroleum, meteorology, and finance.

Bo Shen is a senior engineer of the Inspur-Intel China Parallel Computing Joint Lab, and is engaged in high-performance algorithms, research, and application of software development and optimization. He has many years of experience concerning the development and optimization in life sciences, petroleum, and meteorology.

Guangyong Zhang has a master's degree from Inner Mongolia University, majoring in computer architecture, and is now an R&D engineer of the Inspur-Intel China Parallel Computing Joint Lab, engaged in the development and optimization of GPU/MIC HPC application software. He has published many papers in key conference proceedings and journals.

Xiaowei Lu received a master's degree from Dalian University of Technology, where he studied computer application technology, and is now a senior engineer of the Inspur-Intel China Parallel Computing Joint Lab, engaged in the algorithm transplantation and optimization in many fields. He is experienced in high-performance heterogeneous coordinate computing development.

Qing Wu has a master's degree from Jilin University in Changchun and is now a senior engineer of the Inspur-Intel China Parallel Computing Joint Lab, engaged in high-performance parallel computing algorithm and hardware architecture as well as software development and optimization. He led many transplantation and optimization projects concerning the heterogeneous coordinate computing platform in petroleum.

Yajuan Wang has a master's degree from the Catholic University of Louvain, majoring in artificial intelligence. She is now a senior engineer of the Inspur-Intel China Parallel Computing Joint Lab, and is heavily involved in artificial intelligence and password cracking.

Part I

Fundamental Concepts of MIC

The fundamental concepts of MIC architecture will be introduced in this section, including the development history of HPC, the software and hardware architecture of MIC, the installation and configuration of the MIC system, and the grammar of MIC.

After finishing this section, readers will have learned the background of MIC architecture and how to write HPC programs on the MIC.

High-Performance Computing with MIC

1

In this chapter, we will first review the history in the development of multi- and many-core computers. Then we will give a brief introduction to Intel MIC technology. Finally, we will compare MIC with other HPC technologies, as a background reference of MIC technology for the reader.

Chapter Objectives. From this chapter, you will learn about:

- Developmental history of parallel computing, multi-core and many-core.
- Brief review of MIC technology
- Feature of MIC as compared to other multi-core and many-core technologies.

1.1   A History of the Development of Multi-core and Many-Core Technology

The computer of yore had only one core, and it could only run a single program at a time. Then batch processing was developed about 60 years ago, which allows for multiple programs to launch at the same time. But they could only be *launched* simultaneously; when they run on CPU, they still process sequentially. However, when computer hardware was developed further, one program might not use up all the computational power, thus wasting valuable resources. So the definition of process was born (and then based on processes, threads were later developed). The process switch was also developed, not only making full use of computing power, but also defining the general meaning of "parallel": running different tasks at the same time. However, this "same time" is only based on a macroscopic meaning: only one task can run on a single time segment.

From 1978, after the Intel 8086 processor was released, personal computers became cheaper and more popular. Then, Intel launched the 8087 coprocessor, which was a milestone event (it has great meaning for programmers: the IEEE 754 float standard was born because of the 8087 coprocessor). The coprocessor only assists the main processor, and it has to work together with the central processor.

E. Wang et al., *High-Performance Computing on the Intel® Xeon Phi™*,
DOI 10.1007/978-3-319-06486-4_1, © Springer International Publishing Switzerland 2014

The purpose of the 8087 coprocessor was that at that time, the central processor was designed to work with integers and was weak on float support, but they could not put more transistors into the chip. Thus, the 8087 coprocessor was built to assist with float computation. The significance of this is that the coprocessor was born; the computing process was not patented for CPU, but it now had a first helper. And although after further development of manufacturing technology (i.e., 486DX), a coprocessor was built into the CPU, the idea of a coprocessor never died.

Unfortunately, the growth of computational power never matches our requirements. After the idea of processes was proposed, the computational power came up short again. After Intel announced the 8086 in 1978, CPUs from Intel, AMD, and other companies increased performance continuously, almost following Moore's law: every 18–24 months, the number of transistors doubled, and it allowed for a rapid increase of speed and computational power.

When the manufacturing technology improves, the capability of the processing unit also increases. Technologies such as super-scalar, super-pipeline, Very Long Instruction World (VLIW), SIMD, hyper-threading, and branch prediction were applied to the CPU simultaneously. Those technologies bring instruction-level parallelism (ILP), which is the lowest level of parallelism. By CPU hardware support, they allow parallelism of binary instructions even while running on a single CPU. But this kind of parallelism is commonly controlled by hardware. Programmers can only passively take advantage of technology development instead of controlling the whole process. However, programmers can always adjust their code or use some special assembler instructions to control CPU action indirectly, even though the final implementation is still controlled by hardware.

The developmental speed of CPUs has slowed down in recent years. The primary ways to improve single-core CPU performance are now to increase the working frequency and improve instruction-level parallelism. Both of these methods face problems: while manufacturing technology has improved, the size of the transistor is close to the atomic level, which makes power leaking a serious concern. The power consumption and heat generation per unit size has become larger and larger, making it difficult to improve frequency as quickly as before; 4 GHz is the limit of most companies. On the other hand, there is not much instruction-level parallelism in general-purpose computing. While there is a great effort in changing the design, the performance increase is not proportional to the increase in the number of transistors.

While the single CPU can no longer improve very much, using multiple CPUs at the same time has become the natural next idea for scientists. Using multiple CPUs on a single motherboard is a cost-efficient solution. However, this solution is bounded by cost, so it is only popular on servers, which are not so sensitive to cost and power consumption. The idea of using multiple CPUs is still widely used in the area of high-performance computing.

As early as 1966, Michael Flynn classified computer architecture by instructions and data flow: single instruction stream and single data stream (SISD), single instruction stream and multiple data stream (SIMD), multiple instruction stream and single data stream (MISD), and multiple instruction stream and multiple data stream (MIMD), named as the Flynn classification. Within those classifications,

MISD is very rare, and SISD is referred to as the primal batch machine model. SIMD uses a single instruction controller, dealing with different data streams using the same instructions. SIMD generally describes hardware; it is used in software parallelism, in which the single instruction controller is referred to one instruction instead of the specific hardware. MIMD refers to most of the parallel computers that use different instruction streams to process different data. In commercial parallel machines, MIMD is the most popular and SIMD is the second.

Based on the Flynn's idea of classification for top-level computers, hardware companies are now building supercomputers without taking cost into account. "Supercomputers", referring to those computers with performance in the leading position (e.g., Top500), commonly have thousands of processors and a specially designed memory and I/O system. Their architecture is quite different from personal computers, unlike some personal computers, which are connected to each other. However, there is still a very strong bond between personal computers and supercomputers. Just like high-level military and space technology, which can be used later in normal life (e.g., the Internet), many new technologies used in supercomputers can also be applied to the development of personal computers on desktops. For example, some CPUs of supercomputers can be used directly on personal computers, and some technologies, such as the vectorized unit or the processor package, are already widely used on personal computers.

However, the cost of a supercomputer is too high, so most common research facilities cannot afford it. While network technologies continue to advance, the collaboration of multiple nodes becomes practical. Because all the nodes make a fully functional computer, jobs can be sent to different nodes to achieve parallelism among them, thus using computational resources efficiently. Collaboration can be done through a network, a derivative two-architectures computer cluster, or distributed computing.

A computer cluster is a group of computers connected by a network to form a very tightly collaborating system. Distributed computing is the base of the popular "cloud computing", which cuts a huge computation job and data into many small pieces, distributes them to many computers with a loose connection, and collects the results. Although the performance is no match for the supercomputer, the cost is much lower.

While other hardware companies are extending to different computer architectures, the central processor companies are continuously increasing the frequency of the CPU and changing the CPU architecture. However, limited by manufacturing technologies, materials, and power consumption, after a period of fast development, the CPU frequency has reached a bottleneck, and the progress for increased processor frequency has slowed down. With current methods reaching a dead end, processor companies like Intel and AMD are seeking other ways to increase performance while maintaining or increasing the energy efficiency of processors.

The commercial idea of CPU companies has not changed: if one CPU is not enough, two will be used. Then, with better manufacturing technology, they could put more cores on a single chip; thus, multi-core CPUs were born. In 2005, Intel and AMD formally released dual-core CPUs into the market; in 2009, quad-core and octo-core CPUs were announced by Intel. Multi-core CPUs became so popular that nowadays even normal personal computers have multi-core CPUs, with performance matching the previous multi-CPU server node. Even disregarding the increase of single core performance, multi-core CPUs build multiple cores together, and simply putting them together results in better connectivity between cores than the multi-CPU architecture connected by a mainboard bus. There are also many improvements nowadays for multi-core CPUs, such as the shared L3 cache, so that collaboration between cores is much better.

Along with the transition from the single-core CPU to the multi-core, programmers also began to change ideas, focusing more on multi-threading and parallel programming. This idea was developed in the 1970s and 1980s, with supercomputers and clusters already being built in the high-end of the field at that time. Due to the cost, most programmers could only get their hands on single-core computers. So when multi-core CPUs became popular, programmers started to use tools like MPI and OpenMP, which had been dormant for quite a long time, and could enjoy the full use of computational power.

While the requirements of computational power continue to grow relentlessly, CPU performance is not the only problem. Power consumption is another, so people remembered a good "helper" of the CPU: the coprocessor. In 2007, the popularization of general-purpose computing on graphical processing units (GPGPU) also meant the return of coprocessors. Although the job of GPGPU concerns display and image processing, its powerful float processing capabilities make it a natural coprocessor. And as the developer of the coprocessor, Intel has never forgotten 8087, so in 2012 it introduced the MIC as a new generation of coprocessors, which will make a great contribution to high-performance computation.

There are two key words concerning the development of the computer: need and transmigration. Dire need is always the original power for pushing the development of science and technology; as mentioned above, people needed a faster way to calculate, resulting in the creation of the computer. Then they needed to fully use the resources of the computer, so multi-processing and multi-threading came; because of the increase in calculation capacity, more computation was required, so the processing capacity of one core increased. And as the requirement to the computational power is endless, but improving the processing capacity of one core is limited, different multi-processors came, such as double CPU nodes, clusters, and multi-core CPUs. Transmigration, in general terms, is to recycle. In programmers' jargon it is iteration, the path of computer development, and the way to future development. For example, the primal CPU is in charge of display, but the image processing becomes more and more complex, so the CPU alone cannot handle it. Then the GPU was born, and as manufacturing technology continued to advance, the development of the CPU and GPU became more mature. The CPU and GPU reunited again, such as the i-class CPUs of Intel, the APU of AMD, and NVIDIA is also planning a GPU-supporting arm architecture. Even the coprocessor followed a similar path, from primal 8086/8088, to the single coprocessor, and after manufacturing technology improved, the coprocessor merged with the CPU. But now, due to computational requirements that started from PhysX, the coprocessor was separated again, until the GPGPU was developed as a coprocessor, as well as MIC as a power coprocessor.

Therefore, although people are always talking about how the computer industry develops too fast, it is very easily outdated. But if you can hold the trends and grasp the essence of the problem, you will never be outdated, and can even see the future.

1.2 An Introduction to MIC Technology

Many-integrated core (MIC) architecture, just as its name suggests, is a processor that merges many cores (much more than current CPUs) together. The development code for this particular series is called "Knights", with the aim of leading the trade toward computing a trillion times more than the HPC (high-performance computing) area; it's not a CPU replacement in the computer system, but a coprocessor. A MIC chip usually has dozens of reduced x86 cores, providing high parallel computing capacity. Unlike other coprocessors, the original CPU programs can run on MIC chips as well, which means that we can use the MIC computing resources with only some minor changes to the current programs, thus saving the investment in the original software.

Usually when the MIC series is mentioned, the following words appear: MIC (many integrated core), "Knights" series (such as Knights Corner, KNC), and Intel® Xeon Phi™. MIC is the name of the series' architecture; like CPU, it is the general name of any product using this architecture. Knights series is the code name of the MIC product from Intel; like Ivy Bridge, it is the internal name of one generation of product, and is not used commercially. For example, the first generation uses Knights Corner architecture. When looking at MIC and KNx architecture's object-oriented programming abilities, it is like comparing a father and his kids. Intel® Xeon Phi™ is the series name of a MIC architecture high-performance coprocessor. This name of the MIC series shows Intel's great expectations. The first generation is named Knights Ferry (only developed as a test platform, and not available in the open market), the second Knights Corner, and the third is said to be Knights Landing. As we can see, the knights progressed from the ferry (test), to the battle horn (first generation in the market); will the landing (standing firm in the market), result in "victory"? And of course, will the Knight's Landing in the future be a nice landing?

Compared with Intel multi-core Xeon processors, Intel MIC architecture has a smaller core, more hardware threads, and wider vector units. It is the ideal choice for improving performance and meeting high parallel computing demands. While developers use higher parallelism (instruction, data, job, vector, threads, and cluster), popular and important programming models on Intel processors can extend easily to MIC architecture without changing the programming tools. And the same technologies on Intel processors can provide the best performance. Extending the application to the core, threads, and modularized process data in the multilevel memory and cache can maximize the performance of MIC architecture. By reusing the parallel processor codes, software companies and IT departments of enterprises can take advantage by only maintaining a binary library without retraining their developers for a special programming model related to the accelerator.

The most popular configuration is CPU+GPU heterogeneous computing, where CPU is the mainstay; it mainly focuses on logical operations while GPU runs intensive works. The heterogeneity concerns different instructions. On the other hand, CPU+MIC heterogeneous computing is based on x86 architecture, in which the instruction set in MIC is the same as on the CPU, using the x86 instruction set with some extensions. More importantly, MIC family products are intended to be coprocessors of the Xeon processor family, which makes the program model of MIC the same level as the main processor, working collaboratively; in some circumstances, they can be used as separated computer nodes.

However, the x86 cores contained in the Knights family are different from those in Intel desktop CPUs. The extended instruction set used is similar to the AVX used in Intel's Sandy Bridge, but further extended to 512 bits. Readers with a good memory may still remember the abortive Intel GPU development plan "Larrabee"; Intel now confirms that the mainstay of the MIC project is the Knights family, which inherits most of the achievements of the Larrabee project, with added revisions. It could be called the rebirth of Larrabee, but with a different "heart".

Produced with 22-nanometer manufacturing technology, Knights Corner is the first to support a 3D triple-grid transistor. It will contain more than 50 x86 cores, and the exact number (around 60) will vary with different models.

1.3   Why Does One Choose MIC?

This book explains Intel's solution for high-performance computing and exascale computing, so we shall not focus on any disadvantages of MIC, or make any comparisons with other multi-core and many-core computing solutions, especially with GPGPU (even if MIC is "many-core" in name only, and not "multi-core"). But for users, whether decision makers in the upper level of an enterprise, a user in the middle level, or a programmer in the bottom level, any new solution in the market will have to face the problems of comparison with existing solutions. Even if we can avoid this problem, people will ask, "Why choose MIC? What are the MIC's advantages compared to the current technologies coming from NVIDIA, such as CUDA?"

Each technology must have its advantages and disadvantages compared with current technologies; if it only has advantages and no disadvantages, it is probably only meant to upgrade a program. First, let us briefly review which high-performance computing techniques are now on the market that mainly involve paralleling computing hardware.

1.3.1 SMP

Symmetric multi-processing (SMP) is a kind of multi-processor hardware architecture that has two or more of the same processors sharing the same main memory and that is controlled by one operating system. The most common multi-processor system uses symmetric multi-processing architecture, treating the cores as different processors. Multi-core personal computers that we use and multiple processor servers (a server node with multiple CPU servers) all belong to SMP architecture. In SMP architecture the memory is shared, so it is very easy to use multiple cores, which leads to very high parallelism. Most of the middle and small parallel programs are based on this architecture. When we do parallel programming, threads-level job distribution is used through the OpenMP language extension or pThread threads library.

1.3.2 Cluster

Here, cluster or Beowulf computing generally refers to the structure of a group of computers (usually server-class hardware) connected loosely to form a local network, cooperating tightly to complete computational jobs. "Loosely connected" is in comparison to being on the same circuit board or being connected by a high-speed bus; usually, nodes in a cluster can be used separately, and even if a node is connected to the cluster it can still be easily replaced. "Tight cooperation" refers to the nodes cooperating with each other to complete the same job, and not working individually. One possible example is that if there are N windows in a bank, when it serving M (M > 1) people, the windows cannot be tightly cooperating with each other; when N windows are serving one person (i.e., all windows help one client to count money), it can then be called tight cooperation. Compared to SMP architecture, the connection speed is slower in a cluster; however, it can provide a larger number of parallel processors, as well as more hardware resources (e.g., memory, hard disk). Due to the memory not being shared, normally it has to use message passing, like MPI, to carry out process-level parallelism.

1.3.3 GPGPU

General-purpose computing on graphics processing units (GPGPU) is completing a computing mission that originally should be done on CPUs with graphics processors designed to deal with graphical jobs. Those computations normally have no relation to graphics processing. Due to the high parallelism of modern GPUs and programmable streamline, a steam processor can be used to handle non-graphical jobs, such as solving differential equations. Particularly when dealing with single instruction multiple data (SIMD), for which data processing costs much more time than data organizing and transferring, performance on GPGPU is much better than traditional CPU programs. This is new technology developed in recent years, although GPU has been presented for many years; programming with languages like Cg is still very difficult, and normal programmers have difficulty translating mathematical problems into graphical ones. In 2006, when NVIDIA released the Compute Unified Device Architecture (CUDA), making programming much easier, GPGPU became popular and took off.

NVIDIA's release of CUDA made GPU a software and hardware system of a data-parallelism computing device. CUDA does not require graphical application programming interfaces (API) and uses an easy to handle, C-like language to develop, so there is no need to learn too much new grammar. CUDA architecture includes both hardware and software: the hardware includes CUDA technology (with a core much later than the G80 series) and the software includes a related driver, complier, etc. Both software and hardware must meet the requirements in order to run CUDA technology.

Open Computing Language (OpenCL) is a recent framework for heterogeneous platform programming. This platform can include CPU, GPU, and other types of processors. OpenCL is built with a language (based on C99) for kernels (a function that runs on an OpenCL device), and an API to define and control the platform. OpenCL provides parallelism based on job and data splitting. OpenCL is similar to the other two open standards, Open Graphics Language (OpenGL) and Open Algorithms Language (OpenAL). Those two standards are used in 3D graphics and computer audio. OpenCL extends the capabilities of GPU so it can be used in jobs other than graphics generation. OpenCL is managed by the nonprofit organization Khronos Group. It was first developed and trademarked by Apple and was perfected by cooperating with a technical team made up of AMD, IBM, Intel, and NVIDIA. After that, Apple submitted this draft to Khronos Group. Now, both NVIDIA and AMD GPUs support OpenCL.

GPGPU is very much like the MIC. They are both cost-efficient, high-performance computing solutions compared to CPUs, and even the appearance is similar as both cards use PCI Express (PCI-E) slots. Then what is the difference between the two? And what is the best situation for each card?

First, regardless of "multi-core" or "many-core", the biggest difference is the high parallel computing compared to the original serial codes or codes on multi-core CPUs. This is the same for both MIC and GPGPU. However, they have different definitions for "cores". In GPGPU, using CUDA as an example, core refers to one SP (for the exact definition, refer to CUDA documents); the function of SP is only for computing. In addition, for NVIDIA's Fermi GPU, 32 SPs form an SM, and only two control units are in one SM. In another words, 16 GPU cores must run the same instruction. However, the design idea of MIC is completely different. Every core in MIC can be treated as an x86 core, similar to a CPU on a modern PC with a small server. So programming on MIC can inherit existing parallel codes on the CPU. In some circumstances, MIC cores can even be treated as separated nodes; thus, MIC can be used as a small cluster (this feature will be discussed in detail later). However, x86 cores on MIC have apparently been seriously simplified. If MIC uses the same core as the real CPU core, CPU products numbering at a few dozen or even a few hundreds will be brought out to the market. So although those cores use x86 architecture, they are much stronger than GPGPU cores; please don't expect one of them to be a match for a real CPU, like the Sandy Bridge or the Haswell. MIC relies on the same thing as GPGPU: the huge-crowd strategy. When referring to a "crowd", GPGPU numbers in hundreds, but MIC only has dozens, even on a much more powerful core; how much difference will this make in parallel applications? But don't forget MIC cores are Intel CPUs; although the number of cores can't be much, Intel has a secret weapon: hyper-threading. One physical core can be simulated to be two logical cores. However, every core in MIC can only run four threads at the same time, and those four threads are what Intel calls "hardware threads". They are different from hyper-threads on a CPU since performance is greatly increased, and each thread can be almost treated as a real core. Although hyper-threading on CPU is not as good as single thread, in MIC there is no difference, and most applications on MIC require high parallelism instead of single-core performance; when the parallelism increases, performance will grow. So the "effective cores" on MIC are similar in number to those on GPGPU. Second, MIC uses SMP structure, based on a consistent shared cache, meaning that MIC can use traditional CPU programming models; there is no need to design a new program structure for the new hardware.

MIC only makes little change to existing programs, which can also be shown in easy programming and tools. For easy programming, the offload mode of MIC only adds a few directives that can make programs use MIC to compute. The source code can be shared with traditional CPU programs, reducing the maintenance cost. Tools, such as the traditional Intel compiler, profiler, and debugger, can all be used in MIC programs, reducing the learning cost and saving the initial investment.

People may wonder, if MIC has so many advantages, can it really replace GPGPU? Even though this is a book especially for MIC, we do not believe it will replace GPGPU completely. Although they are similar, thus having a competitive relationship, from a user's standpoint, we need more choice to avoid monopoly. On the other hand, they could also divide up the market, requiring the use of a different product for different requirements, which may be more profitable.

In fact, stronger cores and lighter cores have their own advantages: GPGPU has lighter cores, so its thread switching time can be ignored—this attribute cannot be achieved by the heavier MIC cores. So when programming with GPGPU, using thousands of threads is quite common, and can fit the requirements of GPGPU. On the other hand, for MIC, maybe only hundreds are the maximum. Therefore, generally speaking, GPGPU is suitable for massively paralleled, less branching programs, while MIC is suitable for highly paralleled, logically complicated applications. Here, we mean "simple" logic to refer to less branching or loop instructions.

MIC Hardware and Software Architecture

2

This chapter introduces the hardware and software architecture of MIC. The hardware architecture is the source of the performance, while the software architecture is the foundation of the performance. Deeper understanding of the hardware and software architectures helps programmers to develop and perform tuning efficiently, releasing the maximum potential of the computing system. If an engineer encounters a problem that is difficult to explain in the software development process, he or she may find the answer in the underlying hardware and software architecture.

This chapter is divided into two parts. The first part deals with MIC hardware architecture: from top to bottom, microcosmic to macrocosmic, the core to the periphery—we introduce the MIC hardware features one by one. This part covers the whole MIC and core architecture, including the vector processing unit (VPU), the high-performance on-chip bidirectional ring communication interface, the identical L2 cache, and how to run these functions in interactive mode. We will focus on the key components, such as the performance index and its principle, and basic hardware knowledge of program optimization, such as cache organization and memory access bandwidth.

The second part of this chapter introduces MIC software architecture, which is based on the MIC software hierarchy structure. We'll start from MIC's operating system and drive, and we'll end with API user programming. This part also provides a systematic introduction to the MIC basic software stack, giving the user an exhaustive outline of the MIC software hierarchy structure and the programming principle and method. We hope that this will help the user achieve a deeper understanding of the MIC software ecosystem and develop high-efficiency and high-performance application programs on the MIC architecture platform.

Chapter Objectives. After reading, you should have a deeper understanding of:

- The MIC hardware and software architecture
- The development and optimization theory of MIC, and MIC performance optimization

E. Wang et al., *High-Performance Computing on the Intel® Xeon Phi™*,
DOI 10.1007/978-3-319-06486-4_2, © Springer International Publishing Switzerland 2014

2.1   MIC Hardware Architecture

2.1.1   Definitions

First, some technical terminology:

1. **MIC**. Architecture code. Abbreviation for Intel's Many Integrated Core. Products based on this architecture are called MIC coprocessors; the PCI-E card integrates many MIC architecture cores.
2. **Intel® Xeon Phi™**. Name of the MIC product family, a new member of the Intel® Xeon family.
3. **KNC**. Short for Knights Corner, which is the internal development code for the first generation of Intel® Xeon Phi™ coprocessors.
4. **Host**. The Intel Xeon processor platform that has an Intel MIC architecture coprocessor installed through PCI-E. The host must have the RedHat Enterprise Linux* 6.0 or SuSE* Linux* Enterprise Server SLES 11 or above operating system (Linux Kernel 2.6.32 or above) installed.
5. **Target**. Corresponds with the host, referring to the Intel MIC coprocessor and its related environment when it is running on the host.
6. **µOS**. Operating System based on GNU Linux that runs on the MIC coprocessor. The memory on MIC is separated into two parts: one is controlled by µOS, and the other is controlled by the driver. Driver-controlled means that only the driver can see it, and it is mainly used to transfer data and programs between the host and the MIC. While programs running on MIC use malloc/new to allocate memory, it is distributed from the memory controlled by µOS. The memory controlled by the driver, which cannot be seen by users, can be used by low-level tools.
7. **ISA**. Short for Instruction Set Architecture.
8. **VPU**. Short for Vector Processing Unit.
9. **Ring**. One chip, two-way circling, high-speed interconnection bus that connects the core, memory, and PCI-E together, to achieve high-speed communication.
10. **CRI**. Short for the Core Ring Interface, or the interface between core and ring.
11. **Offload Compiler**. Heterogeneous compiler that can compile binary code to be run on both the host and the MIC. Includes Intel C++ Compiler XE and Intel Fortran Compiler XE.
12. **SDP**. Short for Software Development Platform. It is the development and test platform that includes both the machine supporting the MIC coprocessor and the Intel MIC architecture coprocessor card.

2.1.2   Overview of MIC Hardware Architecture

MIC coprocessors include multiple on chip high-speed MIC architecture computing cores. The MIC coprocessor not only has IA computing cores, but also 8 memory

Fig. 2.1 Extending MIC coprocessor on a traditional Xeon CPU

controllers, which support 16 transferring channels of GDDR5 in total, and can theoretically provide transfer speeds of up to 5.5 GT/s. It also has special-function units, such as the PCI Express system interface.

Figure 2.1 shows a sketch of an extending MIC coprocessor on a traditional Xeon machine.

Every Intelligence Architecture (IA) MIC core is a fully functional, separated, in-order execution IA instruction (IA x86 instruction) core. These cores support hardware hyper-threading, and every core can run instructions from four different processes or threads at the same time. To reduce the memory conflict in the hotspot of a program, MIC has built a distributed tag directory, which lets all physical memory that the MIC coprocessor accesses be linked to this tag directory through a reversible and one-on-one hash function. This hash function not only transfers the entire physical memory to a tag directory, but also provides a consistent framework of the multi-core system, achieving a consistent framework that is much more complex than a single-core system.

A MIC coprocessor has eight memory controllers in total. Each of them meets GDDR5 standards, supports dual-channel memory, and lets the coprocessor attain a transfer speed of up to 5.5 GT/s, so that in theory, data transfer between computing cores and the GDDR5 memory can reach 352 GB/s.

The MIC coprocessor chip is built with the following key components:

1. Computing core:
   (a) Dual-launch, sequential-process-instructed x86 computing core, which supports the EM64T instruction-set.
   (b) 512-bit VPU, which can run in 16*32-bit (integer or float) or 8*64-bit (double) mixed mode. The flexibility of the VPU working mode is shown in Fig. 2.2.

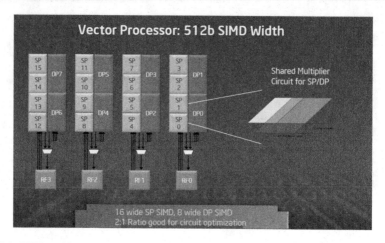

Fig. 2.2 VPU working mode

    (c) Core ring interface (CRI)

       (i) Computing cores and on-chip ring bus interface.

      (ii) L2 cache (including address label, data, status, LRU data), L2 pipe, and related arbitration logical units.

     (iii) Tag directory (TD), which is part of a distributed tag directory.

     (iv) Asynchronous processing of interrupt controller (APIC), which receives interrupts (IPIs, or external interrupt) and redirects the computing core to respond to the interrupt.

2. Memory controller (Gboxes): I/O access to external memory device. Every memory controller has two data-transfer channel control units, and every unit controls a 32-bit access channel.

3. PCI-E terminal logic unit (Sbox): The interface between the MIC system and the host CPU and other PCI-E devices. It supports PCI-E x8 and x16 configurations.

4. Ring: On-chip ring shape interconnection bus that links all functioning units together.

    Figure 2.3 shows the key components of the MIC coprocessor.

    The microstructure of the Knight Corner coprocessor is shown in Fig. 2.4.

    The following section introduces different functional parts of the MIC coprocessor.

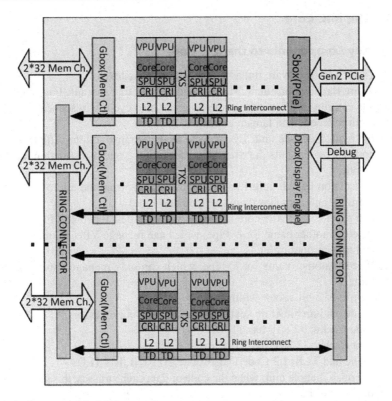

Fig. 2.3 Structure of the MIC coprocessor

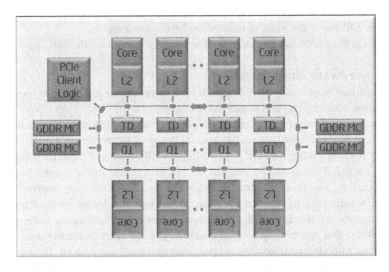

Fig. 2.4 Microstructure of Knight Corner coprocessor

2.1.3   The MIC Core

2.1.3.1 Key Components to the MIC Core

1. T0-T4 IP: four hardware thread units, or thread contexts.
2. Hardware thread controllers: Manage and control four threads. Four hardware threads can run parallel at the same time and also cycling in sequence.
3. Instruction decode and launch units:
   (a) Instruction decode unit: Every cycle reads two instructions (16B) from the code cache and puts the decoded result into the micro-instruction memory uCode.
   (b) Instruction launch unit: Reads two instructions from the micro-instruction memory and launches them to Pipe0 and Pipe1.
4. Launch pipe: Every MIC core has two separate pipes, Pipe0 and Pipe1.
   (a) Pipe0: Another name for U Pipe, which can manage VPU computing units and x86 computing units.
   (b) Pipe1: Another name for V Pipe, which can only manage x86 computing units.
5. VPU Unit: 512-bit vector width VPU.
6. X86 Unit: x86 architecture scalar micro-instruction processor.
7. L1 Code Cache: 32 KB, has TLB.
8. L1 Data Cache: 32 KB, has TLB.
9. L2 Cache and TLB: L2 Code/Data Cache, 512 KB, has TLB.
10. "Cache miss" handle unit: when the code or data cache misses, this unit triggers and deals with the miss.
11. CRI: the link interface between the inner core and the on-chip ring bus.

The MIC core structure is shown in Fig. 2.5.

Figure 2.6 shows the inner details of the MIC core structure.

The following sections discuss different functional units of the MIC core.

2.1.3.2 Hardware Multi-threading

To support hardware multi-threading, the MIC coprocessor has features like basic architecture, stream pipe, and cache inner connection.

Generally speaking, every core has four of the exact same hardware threads; the architecture of those four threads has several complex components: GPRs, ST0-7, fragment register group, CR, DR, EFLAGS, and EIP. Some of the microstructures are also included, such as the preload cache, instruction pointer unit, fragment description flag, and exception treatment logical unit. The following improvements have been imported by the architecture to achieve hardware multi-threading: increased byte of thread ID flag for share architecture resources (like iTLB/dTLB/BTB), and the ability to make memory slots enter thread-related mode through micro-code and hardware support import thread wake/sleep. In summary, the MIC coprocessor achieves an "intelligence" round-robin of multi-threading. Figure 2.7 shows the hardware multi-threading key structure of the MIC coprocessor.

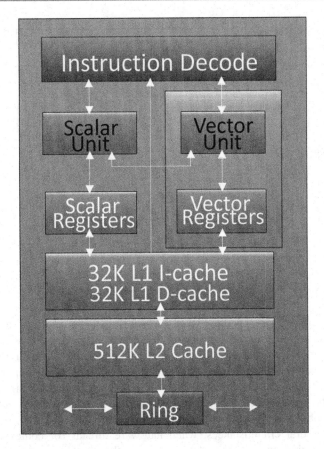

Fig. 2.5 MIC core structure

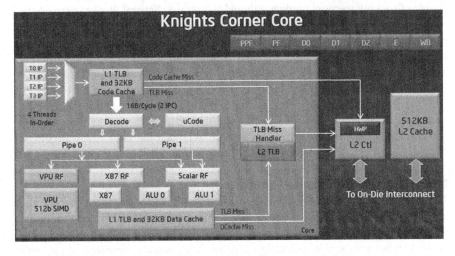

Fig. 2.6 Microstructure of MIC core

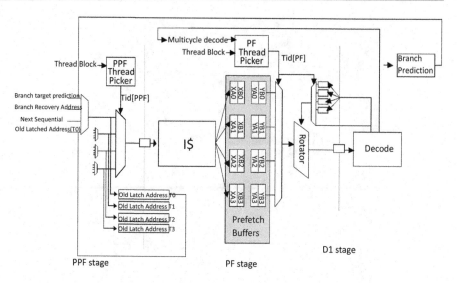

Fig. 2.7 Structure of hardware multi-threading

The gray area of Fig. 2.7 shows that the four threads on a MIC core have one cache, which is "ready to run" and also can do dual-instruction "bundles"; this is called dual-instruction because every core can launch two instructions. If one executing thread needs to jump to an instruction address not included in the buffer, it triggers a "cache miss" event. This event triggers the instruction buffer "refresh context buffer" and loads the objective instructions. If the instruction buffer does not jump to object the instruction, it will halt the computing core and lose performance. In general circumstances, when hardware threads (or contexts) launch instructions during one clock cycle, it preloads the next instruction, following priority. Another important component is the picker function (PF), which is in charge of picking the next hardware thread to execute. Several clocks make up the clock-cycle of the PF. The PF works in a round-robin work mode: in one PF clock-cycle, instructions on only one hardware thread can be launched (one or multiple subsequent instructions rely on what instruction it executes, and different instructions may have different instruction cycles). For example, on clock-cycle N, the PF launches several instructions from Context3. Then, in cycle N+1, the PF will cycle though Context0, Context1, and Context2, and choose one of them for the executing thread/context of this cycle, and then launch the instructions on this context. As mentioned above, it is impossible to launch the same hardware thread (Context3 in this example) continuously (one after another).

2.1.3.3 Instruction Decode/Launch Unit

The decode unit in the MIC coprocessor works on a two clock-cycle, so if the computing core of the MIC coprocessor is working in full stream flow, the same hardware thread/context cannot launch instructions continuously in every clock-cycle. In other words, if it is clock-cycle N, one core launches one instruction from hardware thread1, and then in cycle N+1, this core can launch hardware threads

Core Pipeline

PPF	Thread selector		D2	Micro instruction execution control Address generation Data cache check Register file read
PF	Instruction cache check Preload cache write in		E	Complete ALU execution Confirm exit/delay/exception
D0	Thread selector Instruction cycling Decode 0f,62,D6,REX prefix		WB	Complete register file write Branch code analysis
D1	Instrcution decode CROM check Minimum register unit file read			

Fig. 2.8 Inner core pipeline

other than thread1. This can lead to improving the core frequency and highly increased performance: the MIC coprocessor would score well even when using the single thread benchmark.

Due to the fact that the instruction launcher cannot continuously launch on the same hardware thread, if one core can only run one hardware thread, it can use only 50% of the computing power at most. So maximizing the performance of the MIC coprocessor generally requires at least two hardware threads to be launched on every core.

2.1.3.4 MIC Core Stream Pipe

Every computing core can execute two instructions in every clock-cycle: one though the U Pipe and another though the V Pipe. The V Pipe cannot execute all type of instructions. Its synchronization is controlled by match rules. Vector instructions can only run on the U Pipe.

The structure of the inner core pipe is shown in Fig. 2.8.

The core executes EM64T extended instructions in a similar fashion to the Intel Core 2 architecture processor. The width of the integer register file, data path, and main data transfer bus are all expanded from 32-bit to 64-bit. The number of integer registers increases from 8 to 16. The instruction decode unit has changed accordingly, so that it can decipher REX prefixes and new operation code. Code fragments have added L-bit, to modify the fragment function and protection on the MIC coprocessor. The MIC coprocessor also has four added levels of page tables and RIP-related address-seeking features.

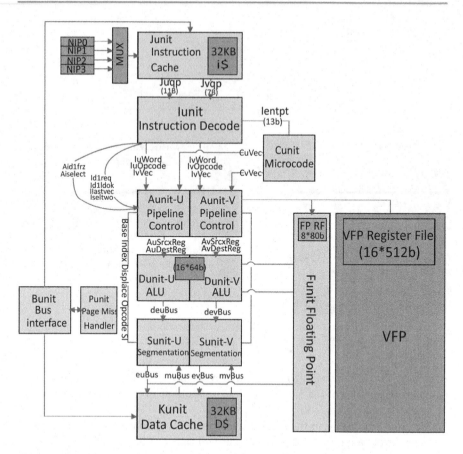

Fig. 2.9 Sketch of MIC core microstructure

The MIC coprocessor further extends the enhanced 64-bit pipeline: computing cores have hardware multi-threading support, reduced performance impact on latency, and the executing unit is kept busy. Every computing core is a 64-bit sequential executing pipeline, which supports four hardware threads. In any clock-cycle, every core can launch two instructions from any single hardware thread/ context as follows: one vector operation instruction and one (special) vector operation instruction or two scalar instructions.

Another sketch of the MIC core microstructure is shown in Fig. 2.9.

Most integer and mask latency is one clock-cycle. Most of the vectorized instructions have a four clock-cycle latency, but executing speed is one instruction per clock-cycle. AGI delay is three clock-cycles. Normal vector operation that is load/store dependent has a latency of four clock-cycles, but shuffles and swizzle address access will increase latency. L1 Cache, if store and load operations access the same physical address, also has a latency of four clock-cycles. The jump access of the data cache will lead to a halt of two clock-cycles because the bank conflict or U Pipe/V Pipe is either replaced by higher priorities or there is a conflict of invalid access. Prefix decode instructions include: "Fast" instruction of zero latency (for example, 62/c4/c5/REX/0f), "Slow" instructions of two clock-cycles latency, "lock", "segment", and "REP". In addition, if the operating number is integer times 66; the address is integer times 67.

2.1.3.5 x86 Architecture Computing Unit

Every MIC core has a x86 architecture scalar processing unit, which can execute standard x86 instructions (for example, EM64T, but not MMX, SSE, or AVX). The x86 computing core can be managed by the U Pipe (Pipe0) and the V Pipe (Pipe1).

2.1.3.6 Vector Processing Unit

The vector processing unit (VPU) is another new component of MIC architecture. Every VPU is linked with a core, and a VPU can only be managed by the U Pipe (Pipe0). A VPU is a SIMD engine built on a 512-bit register. It includes an extended math unit (EMU), so the VPU can execute 16 float operations or 8 double operations; no int-32 or int-64 operations are supported by the VPU of the KNC at this time. 32-float addition and multiplication operations can be executed in one clock-cycle. The sketch of the 512-bit register of a VPU is shown in Fig. 2.10.

The VPU has a vector register file (every hardware thread/context has 32 registers), which can read the operating number (include dynamic format-changing operations) directly from the on-chip memory. Some MIC coprocessor instructions can, through a series of extended instructions, reorganize the data to achieve data broadcast or swizzle. The EMU can execute exp, recip, recipsqrt, log, and pow on SP efficiently, but the DP support is not fully functional.

The SIMD instruction-set VPU that is supported is the Knights Corner instructions (KCi) instruction-set, which doesn't support the traditional vectorized architecture model, such as MMX, SSE, or AVX.

The VPU is designed for best performance on structures of array (SOA) data structures (i.e., [x0, x1, x2, x3, . . . ,x15], [y0, y1, y2, y3, . . . ,y15], [z0, z1, z2, z3, . . . , z15], [w0, w1,w2, w3, . . . ,w15] will be translated to [x0, y0, z0, w0], [x1,y1,z1,w1], [x2, y2, z2, w2], . . . ,[x15, y15, z15, w15] to be processed).

The key features of the VPU of MIC architecture include:

1. Supports double-precision (DP) floating point (FP) ALU.
2. Single-precision (SP) and DP FP ALU both support four kinds of rounding modes, following IEEE754R standards and requirements for rounding on each instruction.
3. It supports SP FP transcendental functions, such as inverse, inverse square root, log2, and exp2. It includes an EMU, which can execute transcendental functions with larger throughput and lower latency.
4. It achieves generic swizzling mux, so it has shuffle/swizzle operating capabilities of 32-bit data size. It gets better performance through two new instructions, VPERMD and VALIGND. It also supports scattered or gathered instructions.
5. The following types of instructions are supported:
   (a) General ISA: vloadunpackh, vpackstoreh, SCATTER/GETHER, GETEXP/PQ, GETMANTP.

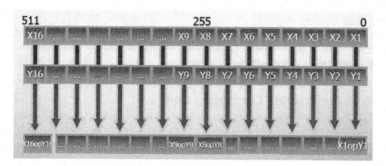

Fig. 2.10 Sketch of the 512-bit register of a VPU

(b) Masked ISA.
(c) Transcend ISA: VRCP23PS, VRSQRT23PS, VLOG2PS, VEXP223PS, VFIXUPNANPD/S, and VFIXUPPS.
(d) Other HPC ISA: VPERMD, VALIGND, VRNDFXPNTPD/S, VCLAMPZPD/S.

2.1.3.7 Core Ring Interface

Each core has a core ring interface (CRI), which is the linked interface between the core and the ring bus on the chip. Its components include:

1. L2 Cache with 512 KB of eight-channel access
2. Queue of all things out-of-core, such as an interrupt signal to the core, request of data access, etc.
3. Data routing logic unit
4. Module R (APIC)
5. Tag directory (TD)

2.1.3.8 MIC Instruction Set

The current instruction set for the MIC coprocessor is KCi (Knights Corner instructions), which is an upgrade of Aubrey Isle.

KCi consists mainly of two parts:

1. EM64T Instruction Set
   The MIC core contains one x86 architecture execution unit (built-in x86 math coprocessor FPU), which is designed based on executing microprocessors in order. It can support the current Intel-64 ISA instruction set EM64T, which can implement 64-bit extensions.
2. Vector Instruction Set for VPU architecture extensions
   Besides having the common vector floating points unit, MIC has a particular 512-bit VPU. The VPU has a bigger register and has a gather/scatter unit (GSU) based on the memory map IO (MMIO), so it can support a mixed selection of the operands and the hardware unit mask. The VPU has 16 general 64-bit registers, but it doesn't support the vectors instruction set of the traditional IA architecture processor, such as the MMX instruction set, the Intel pipelined SIMD extension

instruction set (Intel SSE Instruction Set series), or the Intel AVX instruction set. Therefore, since it is based on EM64T, KCi is a new instruction for MIC architecture.

The list of instruction sets that the MIC coprocessor supports can be found in the Intel MIC Coprocessor Instruction Sets Manual (Knights Corner Instruction Set Reference Manual).

The main features of the MIC Coprocessor Vectors Instruction Set are:

1. New instruction sets for high-performance computing (HPC) applications. These instructions provide the MIC card's local supports for float-32 and int-32 as well as various type conversion supports for multiple common high-performance computing data types. The MIC coprocessor ISA fully supports the arithmetic operations of float-64 and the logical operations of int-64.
2. The core of the MIC coprocessor has now an increased number of registers, viz., 32 new vector registers. The width of each register is 512 bits, and each can contain 16 32-bit float/integer elements or 8 64-bit float/integer elements. Large and continuous vector register documents help produce more effective codes and hide longer instructions delay.
3. Three-operand instructions, which contain two source operands and one target operand. It supports fused multiply and add (FMA): the three operands are all source operands, with one also as the target operand at the same time.
4. Eight vector mask registers, which make the 16 elements of the vector instructions executable depending on the conditions, and saves the mixed result to the original target address. With the mask feature, conditional statements can be used in loop vectorization. In addition, particular vector instructions, such as vcmpps, can also update the value of the vector mask (register).
5. Supports a continuous memory module, which makes the multiple instructions of the new instruction sets able to operate in the same memory address at the same time, just like standard Intel 64 Instruction Sets. This feature makes the development and optimization of vectorization code easier.
6. Introduces some gather/scatter instructions, which can operate on irregular data types in the main memory (through prefetching the memory data in sparse rank to an intensive vector register or reverse operation), using the algorithm for complex data structures to achieve vectorization.

The KCi sets manual contains more information on the MIC coprocessor instruction sets.

2.1.3.9 Cache Organization and Level Structure

1. L1 Cache
   The L1 Cache comprises the 32 KB L1 instruction cache and the 32 KB L1 data cache. It is designed to fulfill the four hardware threads/contexts high-frequency access requirements. The L1 Cache has eight channels, width 64 bytes, bank size 8 bytes. The cache data can be sent back to the core in any order, and the latency is three clock-cycles.

2. L2 Cache

Every core has a globally available 512-KB L2 Cache. It includes eight accessing channels (64 bytes to every channel), 1,024 sets, two banks, 32 GB (35 bit address-seeking) cacheable address memory, and an accessing latency of 11 clock-cycles. Anticipated idle accessing latency is around 80 clock-cycles. The L2 Cache has a stream hardware preload unit, which can selectively preload code, read, and read for ownership (RFO) into the L2 Cache. The L2 Cache has 16 hardware preload streams and can preload 4 KB of data page; if the direction of any stream is set, the preload unit can launch as much as four preload requests at the same time. It supports ECC, as well as new power status, such as C1 power status of core (Close of core or VPU clock), C6 power status (Close of core, VPU clock, or power) or C3 power status of the whole card. The LRU class method is the replacement method for the L1 and L2 caches.

The L2 Cache is part of the core-ring interface, linking the tag directory (TD) and ring stops (RS). These subblocks form the engine of transferring protocol, the interface between cache and RS, which is the same as the frontal bus interface. RS is in charge of managing every message passed to the ring, including the occupation and release of the ring. The TD is distributed physically. It's in charge of filtering ring access requests, and relays the requests to target RS. The TD can also initialize data transfer between the core and GDDR5 memory though an on-chip memory controller.

3. Tag Directory (TD)

The tag directory (TD) can also be called the distributed copy TD because it keeps TD copies of the L2 Cache in every core, and globally all of the L2 Caches of every core. Every time the cache misses, the L2 Cache of every core is kept consistent through the TD. Every TD includes address, status, and the related L2 Cache ID. All TDs divide address memory equally, so when the cache misses, the TD it accessed may not be the one on the core that triggered the cache miss event, but it must be related to the address for the cache miss. The core that has the cache miss event will send an access request to the related TD though the ring.

A sketch of the TD is shown in Fig. 2.11.

4. Introduction of the working principles for the L1/L2 Cache and TD

Intel Pentium is a sequential processor, and cache miss triggers a core event, interrupting the current program it is running, until the required data is loaded and ready in cache. However, when the Intel® Xeon Phi™ coprocessor core has an L1/L2 cache miss, it will not halt the core or the thread that made the access request unless a load miss happens. When any hardware thread/ triggers a load miss, this hardware thread/ will halt until the required data is loaded into cache. However, other hardware threads will continue to run unaffected. Both L1 and L2 Caches can support 38 unfinished read or write requests per core. In addition, system modules (this includes the PCI-E module and the DMA controller module) can generate 128 unfinished read/write requests, so the coprocessor can generate ($38 \times$ number of cores + 128) requests in total. This allows the software to preload data, avoiding any halt related to the data in cache. When all the cache slots are in use, new cache requests will trigger the core to halt until a cache slot is released.

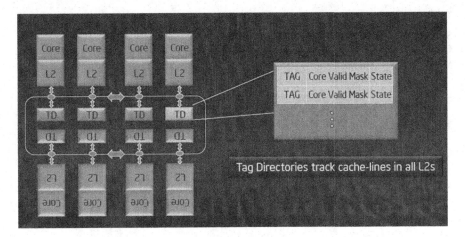

Fig. 2.11 Sketch of the TD

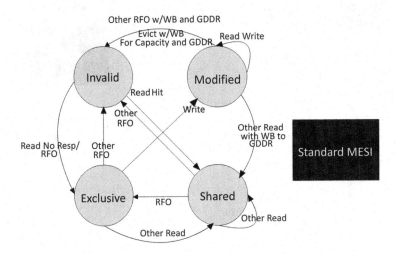

Fig. 2.12 Status of MESI

Both the L2 and L1 Cache use the standard MESI protocol to maintain the shared cache status between cores. Figure 2.12 shows the standard MESI status. Table 2.1 shows the meaning of MESI status.

To locate the bottleneck due to the lack of O status of the MOESI protocol, the MIC coprocessor consistency system has a TD similar to a multi-channel multi-core system. The TD follows the Globally Owned Locally Shared (GOLS3) protocol in addition to the MESI protocols of independent cores. It can simulate an O status miss, achieving full efficiency of MOESI protocol during cache block conditions. The TD can also aid other parts of MIC coprocessor: enhanced MESI and TD GOLS consistency protocols are shown in Figs. 2.13 and 2.14. Table 2.2 shows the meaning of the TD status.

Table 2.1 MESI status

Status of L2 Cache	Definition of status
M	Modified. Cache pipe update according to GDDR Memory. Only one core can have M status cache pipe at the same time
E	Exclusive. Cache is consistent with memory. Only one core can have E status cache pipe at the same time
S	Shared. Cache is shared by cores, and keeps consistency, but may not consistent with main memory. Multiple cores can have S status cache pipe
I	Invalid. Cache pipe is invalid for L1 or L2 cache of current core

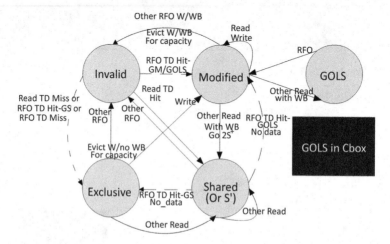

Fig. 2.13 Enhanced MESI statuses

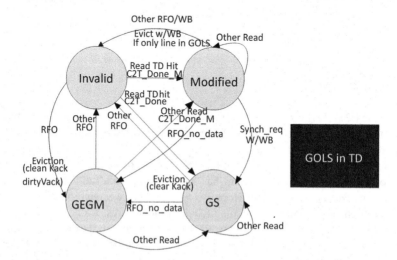

Fig. 2.14 Sketch of TD GOLS consistency protocol

Table 2.2 TD statuses

Tag directory status	Status definition
GOLS	Globally Owned, Locally Shared. Cache pipe is shared by multiple cores, not consistent with memory
GS	Globally Shared. Cache pipe is shared with one or more cores, and consistent with memory
GE/GM	Globally Exclusive/Modified. Cache pipe can be occupied by one core, not sure if consistent with memory. TD is not sure if cache pipe has changed
GI	Globally Invalid. Cache pipe is invalid for any core

The TD is not set up globally; distributed tag directories (DTDs) are used. Every DTD is in charge of maintaining the consistency of the related cache pipe. Related parameters of L1 and L2 cache are shown in Table 2.3.

Table 2.3 Parameters of L1 and L2 cache

Parameter	L1	L2
Coherence	MESI	MESI
Size	32 KB(code) + 32 KB(data)	512 KB
Associativity	8-way	8-way
Line Size	64 bytes	64 bytes
Banks	8	8
Access Time	1 cycle	11 cycles
Policy	LRU like	LLRU like
Duty Cycle	1 per clock	1 per clock
Ports	R or W	R or W

The special items of Table 2.3 are the design of Duty Cycle and Ports; they only appear in the MIC coprocessor. The L1 Cache can be accessed in any clock-cycle, but the L2 Cache can only be accessed every other cycle. In addition, software can read or write to L1/L2 Cache in any clock-cycle, but it cannot read and write at the same time in order to avoid read after write errors.

Every core has a L2 Cache that serves as the second level of cache for instructions and data of the L1 cache. At first glance, it is difficult to understand how huge (up to 31 MB) a globally shared L2 cache must be to guarantee that all cores cooperate with each other. Every core contributes a 512 KB L2 Cache, which is a part of the 31 MB huge "globally shared" L2 Cache, which seems practical. However, if two or more cores share data, the data from the L2 caches of those cores will be completely the same. In other words, if every core does not share data or code with each other, the effective L2 Cache on chip is 31 MB. But if all cores share the same code or data, the effective on-chip L2 Cache is 512 KB. Therefore, the actual effective L2 Cache size is a function of the extent of sharing data or code between cores or threads.

A simple summary for the many-cores of the MIC coprocessor is the chip-level symmetric multi-processor (SMP). It appears that every core is independent of each

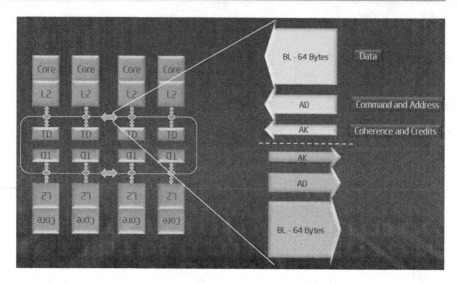

Fig. 2.15 Outline of ring logic

other, with a 512 KB cache. A single chip has many cores (the specific number depends on the model) interconnected by a high-speed inner bus on the chip. If not compared strictly with a real SMP chip, the above simple model can help us to understand how large the available L2 Cache is when the MIC coprocessor is running a specific job.

2.1.4 Ring

The ring includes interfaces between every component of MIC: ring stops, cycling unit of the ring, address-seeking unit, and flow control unit. There are two such rings on the MIC coprocessor: one for each direction. On the ring or one cycle of the ring, there are no queue features. In some circumstances, there isn't enough space to receive the message, so it can be kept on the ring until it is received in the next cycle; this is called a ring jump.

The outline of this logic is shown in Fig. 2.15.

Computing cores and their private L2 Caches interconnect to other units on-chip through the Ring on the CRI link chip. An agent on the ring can be a computing core, an L2 Cache slot, a GDDR memory controller, or an SBOX unit. A pause event can be generated when an occupied or located signal is sent to the ring.

2.1.5 Clock

The clock pulse of the MIC coprocessor provides a clock signal for the four main clocks on the MIC coprocessor:

Core clock: supports from 600 MHz to the maximum frequency, increasing by 25 MHz steps. The frequency of the computing core can be configured seamlessly in real-time with software or built-in hardware (based on thermal and update information from an electrical current sensor).

GDDR accessing frequency: supports 2.8 GT/s to maximum with a 50 MT/s step.

The MIC coprocessor can change working frequency without re-booting.

PCI-E interface frequency: supports PCI-E 1.0 and PCI-E 2.0 working modes.

The MIC coprocessor includes an external clock buffer so that it can work at a related clock-cycle based on two 100 MHz PCI-E frequencies.

2.1.6 Page Tables

The MIC coprocessor supports 32-bit physical address seeking and 36-bit physical address extension (PAE) in 32-bit mode, and supports a 40-bit physical address seeking extension in 64-bit mode.

It supports 4 KB, 64 KB, and 2 MB pages, and can execute disable bit NX. However, it does not support global page tag bits like other IA architecture processors. If TLB misses, the four levels page and INVLPG will start to work; this improvement makes single memory seeking (e.g., 2 MB page table) unrestricted in size (4 KB, 64 KB or 2 MB) of mixed pages within the block, and mixed memory seeking can execute normally. However, if 16 4 KB pages are not consistent with the main memory, it will trigger unpredictable and abnormal behavior.

Every L1 data TLB (dTLB) has 64 4 KB page table entrances, 16 64 KB page table entrances, and 8 2 MB page table entrances. In addition, every core has an instruction TLB. It only has 32 4 KB page table entrances. Instruction TLB does not support a large page table cache. L2 Cache and 4-way dTLB have 64 entrances, so they can be used as a level 2 TLB of 2M memory page or a page directory entry (PDE) of 4 KB/64 KB memory page. If the following registers are the same as multiple threads, TLB can share the page entrances with those threads: CR3, CR0. PG, CR4.PAE, CR4.PSE, or EFER.LMA.

The features of the L1 and L2 Cache are shown in Table 2.4.

Table 2.4 Features of L1 and L2 cache

TLB type	Page size	Page entrances	Structure	Mapping space
L1 Data TLB	4 K	64	4-way	256 K
	64 K	16	4-way	1 M
	2 M	8	4-way	16 M
L1 Instruction TLB	4 K	32	4-way	128 K
L2 TLB	4 K, 64 K, 2 M	64	4-way	128 M

There are two types of memory on MIC coprocessor cores: UnCacheable (UC) and Write-Back (WB). The other three types of memory are: Write-Through (WT), Write-Combining (WC), and Write-Protect (WP). They all map internally to the UC type; no other types are supported.

2.1.7 System Interface

The system interface includes two key components: Knights-corner system interface (KSI) and the transaction control unit (TCU). KSI includes all PCI-E control logic, such as the PCI-E engine protocol, SPI of the on-chip memory, µOS load, I2C fan control, and APIC control logic. TCU bridges KSI and the ring interface of the MIC coprocessor, and also includes the DMA hardware supporting unit and transmission flow control cache. The TCU includes a series of controllers that the DMA requires: code/decode engine, MMIO register, and other queue instructions used for flow control. These instructions provide an internal interface: the on-chip ring transmission protocol engine.

The system interface is made up of many submodules. It can be divided into two functional modules:

1. GBOX

   The memory controller of the MIC coprocessor has three parts: FBOX, MBOX, and PBOX.

   FBOX: on-chip ring interface.

   MBOX: memory access request scheduling unit. It includes two separate CMC (Channel Memory Controllers). MBOX interconnects different modules of the MIC system and links the IO module of DRAM. It links both PBOX and FBOX. Every CMC on the MIC works separately.

   PBOX: physical link to the MIC GDDR5. It is the simulated interface of GBOX, and is in charge of the link with the GDDR memory device. Besides the simulation module, PBOX also includes I/O FIFO, part of the intelligence training unit, and the mode register, which reduces simulated interfaces. Simulated interface is comprised of the actual I/O transmission unit, address seeking unit, instruction/control unit, and clock unit. The PBOX also includes the GPLL module, which defines the clock domain for each MBOX/CBOX of the PBOX.

2. SBOX

   PCI-E logical control unit: DMA engine, and has some power management functionality.

   SBOX is the bridge between the core and external devices. It links the PCI-E bus and the controller of the on-chip thermal sensor.

   The system interface of the MIC card is shown in Fig. 2.16.

   The following sections introduce important interface modules.

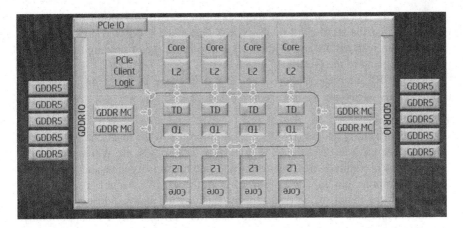

Fig. 2.16 System interface of the MIC card

2.1.7.1 PCI-E Interface

The MIC coprocessor is an expansion card on the PCI-E 2.0 x16 interface, supporting 64-256B data transfer. The MIC coprocessor supports the PCI-E end-to-end read and write functions.

The following two registers show the configuration of an MIC coprocessor: PCIE_PCIE_CAPABILITY Register (SBOX MMIO offset 0x584C)

Bits Type Reset Description
23:20 RO 0x0 Device/Port Type
[other bits unmodified]
PCIE_BAR_ENABLE Register (SBOX MMIO offset 0x5CD4)

Bits Type Reset Description
0 RW 1 MEMBAR0 (Aperture) Enable
1 RW 1 MEMBAR1 (MMIO Registers) Enable
2 RW 0 I/O BAR Enable
3 RW 0 EXPROM BAR Enable
31:4 Rsvd 0

2.1.7.2 Memory Controller

The MIC coprocessor has eight on-chip GDDR5 memory controllers, and every controller can manage two 32-bit memory channels; the transfer speed of each channel can reach 5.5 GT/s. The memory controller can be directly linked to the ring bus so that it can respond to access requests from full physical memory addresses. It is in charge of GDDR memory read and writes, translating memory read and write requests into GDDR commands. All of the access requests from the ring are handled in order, and taking into consideration GDDR access time limitations, there is a great increase in the efficiency of GDDR access, achieving performance close to physical bandwidth. The memory controller in particular guarantees the latency of special requests from the SBOX within the limit. It guarantees that the bandwidth of the SBOX can reach 2 GB/s. The MBOX can communicate with FBOX and PBOX and is also in charge of sending the memory refresh command to the GDDR.

The GDDR5 interface supports two main data check methods: an instruction/address parity check and a software-based data ECC check.

1. DMA unit

Direct memory access (DMA) is a common hardware functionality of computer systems. It releases the CPU from large data copy work. When a copy of a data block is required, the CPU first creates and fills a data cache, and then writes the descriptor into the descriptor ring of the DMA channel. A descriptor describes the detail of the data transfer, such as the source address, objective address, and data size. The following data transferring types are supported by DMA:

(a) Data block transfer between the GDDR5 spaces of the MIC coprocessor A and the MIC coprocessor B.
(b) Data block transfer from the GDDR5 of the MIC coprocessor to the host system memory.
(c) Data block transfer from the host system memory to the GDDR5 of MIC coprocessor.
(d) Data block transfer within the GDDR5 of the same MIC coprocessor.

The DMA descriptor ring is achieved by the μOS on the chip or by coding with the host driver. Eight descriptor rings can be set up through the software so that every ring linked to a DMA channel, μOS, or the host driver can open separate DMA channels on the host system or chip. In other words, the descriptor rings belonging to the host are stored in the host memory, while those belonging to the on-chip μOS are stored in the on-chip GDDR5. A programmable priority arbitration agency is in charge of handling the conflict of multiple DMA channels competing for resources from the host or MIC coprocessor.

In summary, the DMA controller can function as follows:

(a) Support eight DMA channels at the same time. Every channel has a separate hardware ring cache, which can exist on the host or locally.
(b) Support block data passing between the host and the MIC coprocessor, or vice versa.
(c) Support any device launch DMA request, thus supporting any hosts or MIC coprocessor devices in launching a data block transfer.
(d) Always transfer directly with a physical address.
(e) Interruption is generated when the DMA transfer finishes.
(f) 64B granularity alignment and size magnification.
(g) Write a finish mark on the local or host when the DMA transfer finishes.

The DMA module works on the clock frequency of the MIC core. The eight separate channels can pass the following data:

(a) From the GDDR5 memory to the host memory.
(b) From the host memory to the GDDR5 memory.
(c) From the GDDR5 memory to the GDDR5 memory.

Every time the MIC coprocessor transfers data the package size is 64 bytes (1 cache line); the maximum package size a DMA transfer can support is 256 bytes. The actual data package size for each transfer can be set with the MAX_PAYLOAD_SIZE value of the PCI_COMMAND_STATUS register.

2. Interruption handling support

The MIC coprocessor supports the following three interruption types:

(a) Local interrupts: Interrupts with objects are cores on the "local" MIC coprocessor. These interrupts show as APIC messages on the APIC serial bus.

(b) Remote interrupts: Interrupts with objects are cores on the "other" MIC coprocessor. These interrupts show as MMIO access requests on the PEG port.

(c) System interrupts: Interrupts with objects are on the host processor. These interrupts show as INTx/MSI/MSI-X messages on the PEG port, depending on the PCI configuration.

2.1.7.3 Memory Space of MIC Coprocessor

The distribution of memory space for the MIC coprocessor is shown in Table 2.5.

Table 2.5 Distribution of memory space of MIC coprocessor

Function	Start address	Size (bytes)	Remark
GDDR5 Memory	00_0000_0000	Variable	–
System Memory	–	Variable	Mapped to host system address through system memory page table (SMPT)
Flash Memory	00_FFF8_5000	364 K	Actual size of flash memory; parts of it cannot be accessed through the normal accessing path
MMIO Registers	00_007D_0000	64 K	Access feature related to host side
Boot ROM	00_FFFF_0000	64 K	New additional memory of the MIC coprocessor, with FBOOT0 image in flash memory
Fuse Block	00_FFF8_4000	4 K	New additional memory of the MIC coprocessor

1. Memory space, visible on host, of the MIC coprocessor

After restarting, the GDDR5 memory of the MIC coprocessor enters the "stolen memory" mode. In this mode, the on-chip GDDR5 is not visible or accessible to the host. Stolen memory mode (CP_MEM_BASE/TOP) has higher priority than the PCI-E aperture mode. Stolen memory can be reduced or even removed by configuring the FBOOT1 code. PCI-E aperture mode can create massive memory space through host programming or distribution on the μOS.

2. Boot section of the MIC coprocessor

The boot process of the MIC coprocessor can be summarized as the following:

(a) After reboot: Boot strap processor (BSP) directly executes the code in the boot image (FBOOT0) stored in the first stage.

(b) FBOOT0 certifies the boot launcher (FBOOT1) stored in the second stage and jumps to FBOOT1.

(c) FBOOT1 launches/trains the GDDR5 basic memory mapping table.

(d) FBOOT1 tells the host to upload the μOS image onto the GDDR5.

(e) FBOOT1 certifies the μOS image; if it fails, FBOOT1 locks some of the special functions of the μOS.

(f) FBOOT1 jumps to the μOS.

3. SBOX MMIO registers

The SBOX includes 666 memory mapped I/O (MMIO) registers, with a total size of 12 KB. These registers are used to configure the function of the MIC coprocessor, record the statuses of the MIC coprocessor, and debug the SBOX or other functions of the MIC coprocessor. Therefore, those registers are sometimes called configuration and status registers (CSR); however, the PCI-E configuration registers are not included.

The address range of the SBOX MMIO register groups is 08_007D_0000h–08007D_FFFFh. MMIO register groups are not continuous, but are separated by the function unit of the SBOX.

The SBOX MMIO register groups are also accessible by the μOS. However, due to safety reasons, the host access is limited.

4. Host and MIC architecture physical memory mapping

The mapping between physical addresses of the host and MIC coprocessor is shown in Fig. 2.17.

The MIC coprocessor memory supports 40-bit physical addresses and has a maximum memory space of 1,024 GB. This memory space is divided into three higher-level address segments:

1. Local address segment: 0x00_0000_0000 to 0x0F_FFFF_FFFF (64 GB)
2. Reserved address segment: 0x10_0000_0000 to 0x7F_FFFF_FFFF (448 GB)
3. System (Host) address segment: 0x80_0000_0000 to 0xFF_FFFF_FFFF (512 GB)

The local address segment is further divided into four equal address segments:

1. 0x00_0000_0000 to 0x03_FFFF_FFFF (16 GB):
   (a) GDDR (Low) memory segment.
   (b) Local   APIC   segment   (variable   size),   0x00_FEE0_0000   to   0x00_FEE0_0FFF (4 KiB).
   (c) Boot code (Flash) and Fuse, 0x00_FF00_0000 to 0x00_FFFF_FFFF (16 MiB).
2. 0x04_0000_0000 to 0x07_FFFF_FFFF (16 GB)
   GDDR memory (size can reach PHY_GDDR_TOP)
3. 0x08_0000_0000 to 0x0B_FFFF_FFFF (16 GB)
   (a) Memory mapping registers.
   (b) DBOX registers 0x08_007C_0000 to 0x08_007C_FFFF (64 KiB).
   (c) SBOX registers 0x08_007D_0000 to 0x08_007D_FFFF (64 KiB).
4. 0x0C_0000_0000 to 0x0F_FFFF_FFFF (16 GB)

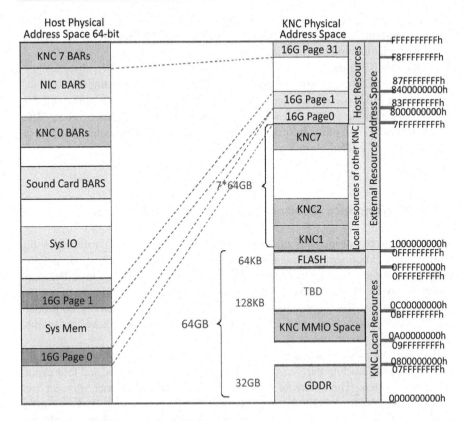

Fig. 2.17 Mapping between physical addresses

System address segment includes 32 memory "pages" with 16GB for each page:

1. Sys0: 0x80_0000_0000 to 0x83_FFFF_FFFF (16 GB)
2. Sys1: 0x84_0000_0000 to 0x87_FFFF_FFFF (16 GB)
3. ...
4. Sys31: 0xFC_0000_0000 to 0xFF_FFFF_FFFF (16 GB)

   The system address segment is used to access host physical memory address, and can "see" a memory space of up to 512 GB. Remote MIC coprocessor devices can also access through the system address. All addresses of requests accessing the Host through PCI-E belong in this address segment. The system memory page table (SMPT) maps the 40-bit local address into the 64-bit host system address.

   If the no-snoop bit of the SMPT register is not set, the host will monitor the access of the MIC coprocessor to host memory. The System Control Interface (SCIF) driver will not set the control bit mentioned above, so in SCIF mode, MIC coprocessor access to the host system memory will always be monitored. In addition, all host access to the MIC coprocessor cache will always be monitored.

The mapping from the MIC coprocessor address to the host address is configured through registers MEMBAR0 and MEMBAR1 (refer to related Intel official documents for more details).

5. Multiple MIC cards

Every MIC coprocessor can be treated as an independent computer system. When the MIC coprocessor starts, the host OS will list the information of all the MIC cards on the system and start separate μOSs and SCIF drivers on all MIC coprocessors. For more information on multiple card communication, please refer to SCIF related chapters.

2.1.8   Performance Monitoring Unit and Event Manager

The MIC coprocessor includes a performance monitoring unit (PMU) similar to the Intel Pentium processor. Although the PMU interface is updated to be compatible with future programming interfaces, it still covers most of the 42 Intel Pentium processor event types. Some of the events centered on the coprocessor are introduced to the MIC coprocessor, and are used to measure memory controller events, vector processing unit efficiency, local and remote cache read/write statistics, etc.

The MIC coprocessor supports an independent single-core performance monitor. Every MIC core has four performance monitors, four filter counters, and four event selection registers. The events of the MIC coprocessor performance monitor are made by the events inherited from the Intel Pentium processor and any new MIC-owned events. The PMU of every MIC core is shared by all four hardware threads. The PMU of every MIC core is in charge of monitoring the time-stamp counter (TSC) on the core and hardware events generated by this core, and is triggered by any events that reach this core. By default, the event counter will count the events generated by all four hardware threads. However, through filter configuration, it can only count specific hardware threads. The PMU inside the MIC core can also receive and count events generated by neighboring parts, including the RS (Ring Stop), the DTT (Distributed Tag Table), and the ring.

The interface of the MIC coprocessor PMU is similar to that of the Intel Pentium processor, allowing user-level (ring 3) programs to interact directly with the PMU and get data collected by the PMU through a specific command (such as RDPMC4). In this mode, although the lowest level is still in control of the PMU, user-level (ring 3) applications can interact with the exposed features of the PMU. This opens a window to the performance of the MIC coprocessor for the developers to focus on performance tweaking.

Please refer to the Intel official manual for further details of PMU commands that control and inquire to the MIC core.

2.1.9 Power Management

Power consumption is always a very important indication of HPC computing; low power consumption is always a key point for chip design. One of the highlights of MIC is its low power consumption design. The power management of the MIC coprocessor supports Turbo Mode and Package State.

Turbo mode is in real time and is dynamically launched. It can increase power output by the number of active computing cores and the workload. It can therefore support higher working frequency and voltage to get higher performance. In contrast to the traditional Intel Xeon processor, the MIC coprocessor does not have a hardware power management unit; its power is managed by the μOS on the coprocessor.

When the workload on the MIC coprocessor is relatively low, the idle computing units can be set to low-cost halt states or even completely shut down to save power.

A brief list of different operating modes and power states is below:

1. CoreC1 State: Single core and VPU clock shutdown (All four hardware threads are halted)
2. CoreC6 State: Single core and VPU power shutdown (C1 state and shutdown timer)
3. PackageC3 State:
   (a) All cores clock and power shutdown.
   (b) Ring and noncore unit clock shutdown.
4. All C6 State: VccP, core, ring, and noncore units shutdown.
5. Memory states:
   (a) M1: Clock shutdown.
   (b) M2: GDDR self-refresh.
   (c) M3: M2 and power down.
6. GMClk PLL
7. SBOX States:
   L1: L1 linking state of PCI-E, SBOX clock closed.

2.2 Software Architecture of MIC

2.2.1 Overview

The software architecture of MIC products is designed for massively parallel applications. MIC products are designed for a card using PCI Express (PCI-E) interface. Therefore, MIC must follow any related PCI-E standards.

From the software angle, every MIC card can be treated as a symmetric multiprocessing (SMP) computing domain. It is loosely coupled with the host domain running the OS. Loose coupling is used so the MIC card can use the same hardware system to support many different programming models. Highly parallel applications generally use different programming models. Therefore, the MIC card supports users who wish to maintain their own programming models as

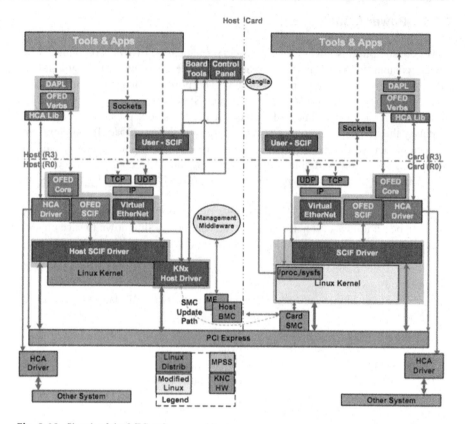

Fig. 2.18 Sketch of the MIC software architecture

much as possible, instead of changing the programming model as a result of hardware design.

To make widely supported tools and applications available in the HPC market, the software stack supports many application programming interfaces (APIs) including standard APIs, such as communication interfaces like TCP/IP and MPI. Some Intel private APIs are also supported. The symmetric communication interface (SCIF) provides a general transmutation API through the PCI-E hardware for other interfaces to transfer data between the host and MIC devices. Figure 2.18 explains all the APIs on the Intel MIC architecture Platform Software Stack (MPSS) and their relationships with each other. To highlight parallel programming, this figure also shows which bottom-level API is needed by the executing file or executing environment.

The left side of Fig. 2.18 shows the levels of the standard Linux core software stack on the host. The right side shows the levels on the MIC. The MIC is based on a Linux core, with some changes.

This figure describes the normal runtime states after a boot through BIOS. The firmware of every MIC card is called "Bootstrap", which will start when the system reboots. Bootstrap configures hardware on-board, and then waits for instructions from the host driver. At the same time, the on-board μOS and other software stacks are loaded, finishing the configuration of the whole software stack setup.

The purpose of software architecture is to support application models at the bottom level. The MIC is specifically in charge of on-board threads and memories. Although it is common to use the host OS to manage all of this, by using this architecture it is possible to call the functions of one device from any other; for example, you can notify the CPU from the MIC side.

To support application models, the software architecture must support communications between the host and MIC devices. This is achieved through the SCIF driver or advanced API built above the higher-level SCIF. The SCIF is designed to reduce latency, achieve low-cost communication, and provide independent communication flow between SCIF clients.

A simulated network interface is located on top of the SCIF ring 0 drivers. It creates the network between MIC devices and the host base on IP. This network can be configured by the host or user to bridge additional networks.

Developers can take advantage of this environment and architecture to build useful models on SCIF. For example, the MPI stacks built on the SCIF can support any MPI message pass model (such as send/receive, one side, and double sides).

From this figure, we also can see that the PCI-E links the MIC card with the host. On the MIC side, it is based on the Linux kernel. Therefore, the Linux system function can be called in programs. On level ring 3 (user level), the programmer can use a traditional socket or higher-level network communication API, such as MPI, to communicate with the host through simulated TCP/IP software stacks on level ring 0. In addition, the programmer can also make use of the user SCIF (note that this is not the system-level SCIF driver, but a user-level API, similar to the user API and system API of the OS concept), using the bottom level to communicate with the host.

In this section, we explain the MIC card boot and on-board OS, and the SCIF driver and system monitoring. Last, we discuss the programming interface related to the network.

2.2.2   Bootstrap

Because the MIC core is an x86 architecture core, the self-check of Bootstrap is similar to BIOS. Bootstrap starts when the card first powers up, but the host will also restart it after a catastrophic failure. The responsibility of Bootstrap is to initialize the MIC card and start the μOS.

Bootstrap comprises two different code blocks: FBOOT0 and FBOOT1. FBOOT0 belongs to the on-chip ROM and is not upgradeable. FBOOT1 is on Flash and is upgradable.

2.2.2.1 FBOOT

After the card resets, the first instruction it runs is on FBOOT0. This code block is the root of trust because it cannot be modified. It is used for verifying the second block, FBOOT1, and transferring the root of trust to FBOOT1. If verification fails, the FBOOT0 will cut off power for the ring and core to prevent any other action. The only way to recover from this state is to change the on-board jumper manually, changing the card to zombie mode. This mode allows the host to reprogram the Flash chip and recover the damaged FBOOT1 code block.

The FBOOT0 running process is the following:

1. Set Caches as RAM (CAR) mode to reduce executing time.
2. Switch to 64-bit protective mode.
3. Verify FBOOT1.
4. If fail, close the card.
5. If succeed, pass the control to FBOOT1.

2.2.2.2 FBOOT1

The responsibility of FBOOT1 is to configure the card and start the μOS. Configuring the card includes initializing all cores, noncore units, and the memory. This process is similar to that of classical x86 cores. The code must be run in 64-bit protective mode to gain access to the required registers.

When starting a third-party μOS (i.e., a Linux μOS based on MPSS), the root of trust is not passed down. It only passes down in maintenance mode, and even then it requires privileged operations. Maintenance mode locks some registers and rewrites some hardware.

After verification is complete, it decides which type of μOS to boot. FBOOT1 calls back FBOOT0, and FBOOT0 uses the embedded public key to run the verification process. In maintenance mode, the μOS signs a private key, while in other modes an unsigned one is used. If verification succeeds, the maintenance mode for the μOS will be launched. If verification fails, it defaults to the third-party μOS and Linux boot protocols and locks access to sensitive registers in order to protect intellectual property.

The FBOOT1 running process:

1. Set the memory frequency and reset card.
2. Finish core initialization.
3. Initialize GDDR5 memory.
   (a) If memory has been trained, use training parameters to recover Flash.
   (b) If there are no recoverable training parameters, or those parameters do not match the current configuration, it will use the normal training procedure and restore the parameters to Flash.
4. Map FBOOT1 to the GDDR5 memory to improve execution time.
5. Initialize uncore parts.
6. Complete CC6 register initialization.
7. Start APs.
8. APs switch to 64-bit protective mode.
9. APs finish core initialization.

10. APs finish CC6 register initialization.
11. APs finish and wait for further instructions.
12. Wait for the μOS to be downloaded from the host.
13. Verify the μOS (all cores participate the verification to reduce time).
14. If verification succeeds, maintain the μOS and launch it.
15. If verification fails, use third-party μOS instead.
    (a) Lock access to registers.
    (b) Create boot parameter structure.
    (c) Switch to 32-bit protective mode, close paging.
    (d) Hand over control to the μOS.

2.2.3 Linux Loader

MIC starts the μOS image based on Linux. It also supports a third-party μOS. Bootstrap will follow some written standards of Linux kernel to start the μOS. There are three potential kernel entry points: 16-bit, 32-bit, and 64-bit. Every entry point adds more and more data structures to configure. Knights Corner uses the 32-bit mode entry point.

2.2.4 μOS

The base running environment made by the MIC μOS is the base of other software stacks. The MIC μOS is based on standard Linux kernel source code. The goal is to change the standard kernel as little as possible in order to reduce maintenance cost (following the GPL license) when migrating to new kernel versions. Although modifications for specific architectures are rare, in some circumstances increasing the changes to kernel is necessary, especially when the standard kernel adds some hardware-related functions that aren't supported on MIC.

As shown in Fig. 2.19, the μOS provides some classical capabilities such as process and task creation, time sequence arrangements, and memory management. μOS also provides configuration, power, and server management capabilities. MIC's Linux kernel can be extended through loadable kernel modules (LKMs). LKMs can be loaded and unloaded by the modprobe command. Those modules may include Intel-provided modules (such as idb server side or performance collector) and end user-provided modules.

The MIC based on Linux μOS is minimized since the embedded Linux migrated to the MIC architecture through the Linux Standard Base (LSB) kernel library. The LSB is also an unsigned OS that uses the minimized shell environment BusyBox. The following are the elements of LSB:

1. glibc: GNU C standard library
    (a) libc: C standard library
    (b) libm: math library
    (c) libdl: dynamic library load on run-time

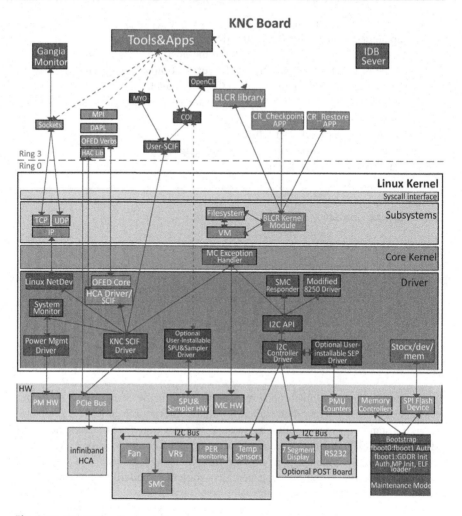

Fig. 2.19 The MIC software stack

    (d) librt: POSIX run-time library

    (e) libcrypt: code and decode library

    (f) libutil: utility function library

2. libstdc++: GNU C++ standard library

    (a) libgcc_s: low-level run-time library

    (b) libz: no-loss data compression library

    (c) libcurses: terminal-unrelated character display library

    (d) libpam: pluggable authentication module (PAM) interface, which allows
        applications to make inquiries in a customized way defined by the adminis-
        trator (separate implementation of program with authentication)

About CPUID

CPUID can acquire some information from the CPU (the processor on MIC is also treated as a CPU), but CPUID on MIC cards does not follow the standards of Intel 64 Architecture Processor Topology Enumeration. However, it can get some information through /sys/devices/system/cpu/cpu*/topology/*.

2.2.5 Symmetric Communication Interface

The SCIF is the base of communication between the host and the MIC heterogeneous computing environment. Its purpose is to provide communication within the same platform. The SCIF not only communicates between the host and the MIC card, but also between MIC cards. It provides a unified communication API through the PCI-E system bus, making full use of the PCI-E hardware transmission capability. The SCIF directly exposes the DMA capability that makes MIC capable of high bandwidth and large-quantity data transmission, thus mapping memory between host and MIC devices, and handling memory address.

SCIF communication between node pairs is based on point-to-point direct physical memory access within the nodes. A special case is when both nodes are MIC cards, because then it does not use the system memory.

The SCIF message layer uses the inherited reliability of PCI-E. It is treated as a simple network that only transfers data without the need for internal package checking. The messages are not numbered, and there are no error checks. Since it only transfers data, it does not directly replace high-level transferring APIs, but provides an abstractive layer to use system hardware for other APIs instead. Every API can use the SCIF to build its own interface.

For more details about the SCIF, please refer to Sect. 5.5, which discusses SCIF programming.

2.2.6 Host Driver

The host driver (KNx host driver in Fig. 2.18) is the collection of host-side drivers and services shown in Fig. 2.20. It includes the SCIF, power management, the remote access system (RAS), and service management. The first job of the host driver is to initialize the MIC card, which includes loading the μOS and setting up the boot parameters for every card. After booting successfully, the primary responsibility of the host driver is to be the root of the SCIF network. Extra tasks are to serve host-side power management, device management, and the configuration interface. However, the host drive does not directly support any kind of user interface or remote API. These tasks are achieved by employing user-level applications or communication protocols built on the driver or the SCIF.

DMA supports asynchronous operations. Adding in DMA on the host has less latency than setting up a DMA agent on the card. The application can choose a

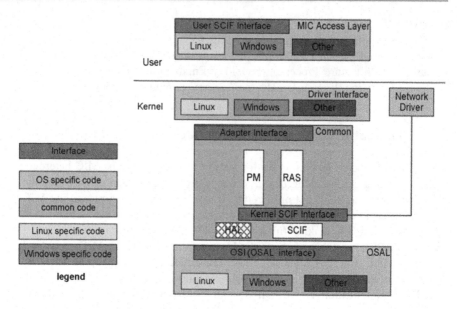

Fig. 2.20 Sketch of the MIC host driver software architecture

memory copy (without using DMA) or let the driver decide the best way. Memory copy is optimized by multi-threading, which can use multiple cores operating in parallel to reach the bandwidth of the PCI-E. However, when low CPU usage is required, or the size of the data transfer is greater than some limitation (the actual limitation has to be tested to confirm), DMA is the best choice.

There are many benefits to the host driver in supporting MSI/x interrupts:

1. No need for specific hardware interrupt link.
2. Does not share interrupt with other devices.
3. Due to optimized hardware design, no need to read back from hardware, thus improving the interrupt handling efficiency.
4. In multi-core systems, this device can trigger interrupts for different cores.

2.2.6.1 Control Panel

The control panel is placed on the host side, providing basic surveillance of the on-card system. The control panel is only suitable for small workstations, not larger clusters. It can provide the following functionalities:

1. Monitor the states, parameters, power, etc., of the MIC card.
2. Monitor system performance, core usage, memory usage, etc.
3. Monitor system health, key failures, and events.
4. Hardware configuration and setup, ECC, etc.

2.2.6.2 Ganglia Support

Ganglia is a scalable, distributed, high-performance application (e.g., network or cluster) surveillance system. Ganglia is robust and has been integrated into many

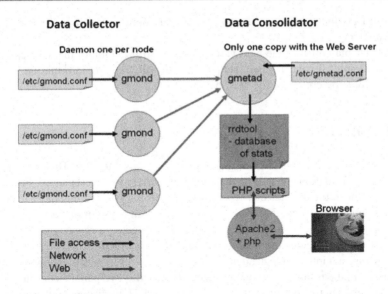

Fig. 2.21 Ganglia data flow chart

operating systems and processor architectures; thousands of clusters are using this system.

To put it simply, Ganglia has daemons on every node or machine; data is collected by those daemons and put into a round robin database tool (rrdtool) database. Ganglia uses PHP to build a web server, showing surveillance data by graphical way. The classic flow chart of Ganglia is displayed in Fig. 2.21.

MIC provides support to Ganglia through the TCP/IP protocol built on the SCIF. In addition to the default collected information, Ganglia can also monitor the following information on the MIC:

1. Use of MIC device
2. Use of on-board memory
3. Use of on-board cores
4. Core temperature
5. Card temperature
6. Core frequency
7. Memory frequency
8. Core voltage
9. Memory voltage
10. Power consumption
11. Fan speed
12. Number of active cores (physical cores)

2.2.6.3 MIC Architecture Commands

In this section, a few MIC commands are introduced. For more detailed usage, please refer to the "–help" of each command.

micflash: Update the MIC card firmware; different options can display some Flash
parameters.

micinfo: Display most of the hardware and driver information of the MIC card.

micsmc: Graphical control panel, surveillance of MIC running information.

micctrl: Boot and configure the MIC card.

2.2.7   Sysfs Node

Sysfs is the virtual file system based on the memory on Linux 2.6. This file system
can output information concerning the device and driver to the user space. For MIC
devices, some of the features can also be obtained from sysfs. Getting these features
(such as core/CPU usage, threads/processes details, and system memory usage) is
better than using the standard /proc interface. The directory organization structure
is a strict internal organization based on the core data structure.

Sysfs has a kind of mechanism for expressing core objects, showing their
attributes and their relations to other objects. It provides two parts: a core interface
to output those items through sysfs and a user interface to display and control the
expressions mapping those items back to the core. Table 2.6 shows the mapping
relation between the two (core and user space):

Table 2.6 Core and user
space mapping relations

Internal	External
Core objects	Directories
Object attributes	Regulation files
Object relations	Symbol links

The following sysfs nodes are available at present:

clst: Number of known cores at present

fan: Fan state

freq: Core frequency

gddr: GDDR device information

gfreq: GDDR frequency

gvolt: GDDR voltage

hwinf: Hardware information (version, step, etc.)

temp: Read thermal sensor

vers: Version string

volt: Core voltage

sysfs: Part of the core kernel, it provides a relatively simple interface to execute a
simple test. Some popular system surveillance software (such as Ganglia) use /
proc or the sysfs interface to get information on system states. MIC uses sysfs to
obtain card information, making the local and server managements use the same
interface.

2.2.8   MIC Software Stack of MPI Applications

This section includes software stacks required by MPI, such as μDAPL and IB verbs. Due to the extensive usage of MPI in the high-performance computation field, MIC has built-in support for the Open Fabrics Enterprise Distribution (OFED). OFED is widely used in the HPC field due to the microsecond level latency. By using the remote direct memory access (RDMA) capability, OFED became the first choice communication stack of the Intel MPI library on MIC. Based on the OFED Intel MPI library on MIC, SCIF or physical InfiniBand (IB) host channel adaptor (HCA) can be used to communicate between the host and the MIC device. In this mode, the MIC device is treated as an independent node of the MPI network. InfiniBand is an I/O system based on channels and using switch structures. It is based on a long cable connection and has high speed and low latency features. InfiniBand uses the internal link between the processor and other parts of the servers (like PCI-E). Now, it commonly uses the link between server nodes (replacing Ethernet). MIC uses InfiniBand-related hardware on MIC for the this reason. MIC communication has two ways to achieve MIC-Direct or OFED/SCIF.

1. MIC-Direct: A proxy driver, allow access to InfiniBand HCA hardware through the MIC device.
2. OFED/SCIF: Software-based InfiniBand-like device, which can communicate within a server node (between CPU and MIC, or between MIC and MIC).

InfiniBand HCA can make both supports within the node or between nodes communications.

2.2.8.1 MIC-Direct
In order to communicate effectively with the remote system, programs running on the Intel MIC coprocessor need to have direct access to host-side RDMA devices. This section explains an internode communication architecture called MIC-Direct.

In a heterogeneous computing environment, whether for CPU processors or MIC coprocessors, a high-efficiency communication solution for all the processors is desired. Providing a general and standard programming and communication model for cluster system applications is an important goal for MIC software. A consistent model not only simplifies the application development and maintenance process, but also provides more agility for better performance.

RDMA architecture (such as the InfiniBand) enjoys great performance success on HPC cluster applications by reducing message-passing latency and increasing bandwidth. The ways RDMA increases performance are moving the network interface to make it closer to the application and allowing kernel bypass, direct data placement, and more control power to fit the I/O request of applications. RDMA architecture allows using hardware to isolate process, protect, and translate address. These features are suitable for the MIC coprocessor environment because programs on the host and coprocessor sides have their own address ranges.

MIC-Direct gives MIC the benefit of RDMA. Without MIC-Direct, communicating inside or between additional processors (such as the coprocessor) results in additional cost to bring data to the host memory, thus greatly affecting in a

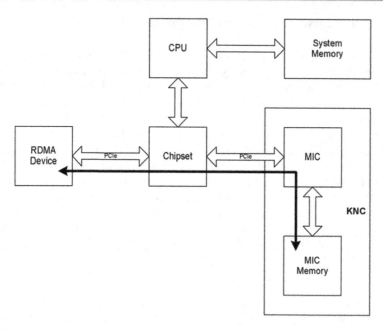

Fig. 2.22 RDMA data transfer pass of MIC-Direct

nontrivial way the communication latency and bandwidth. Figure 2.22 shows MIC coprocessor RDMA transmission attached to PCI-E through MIC-Direct.

MIC-Direct allows applications based on Linux hosts and MIC coprocessors to share RDMA hardware devices. Figure 2.23 is the MPI application that uses MIC-Direct.

Figure 2.23 highlights that the key software modules (bold rounded rectangle) controls MIC-Direct. These modules include the IB proxy daemon, IB proxy server, IB proxy client, vendor proxy drivers, and SCIF. The host system can get one or more RDMA devices and one or more MIC coprocessors through one PCI Express interface. Host-side and MIC software modules communicate with each other, accessing RDMA devices through a PCI-E bus. Software use divides the driver model proxy pass through PCI-E in order to manage the operations on RDMA devices. These operations create resources on the host side through vendor drivers.

1. IB proxy daemon

   The IB proxy daemon is a host-side user-mode application. It provides user-mode context to call a bottom-level vendor driver for the IB proxy server side. A user-mode process context needs to finish virtual address mapping for the RDMA device memory without changing an existing vendor driver. Through udev rules, it creates an instance of an IB proxy daemon using an IB proxy server for every MIC connection.

2. IB proxy server

   The IB proxy server is a kernel module on the host side. It provides communication and commands for the MIC IB proxy daemon. The IB proxy server

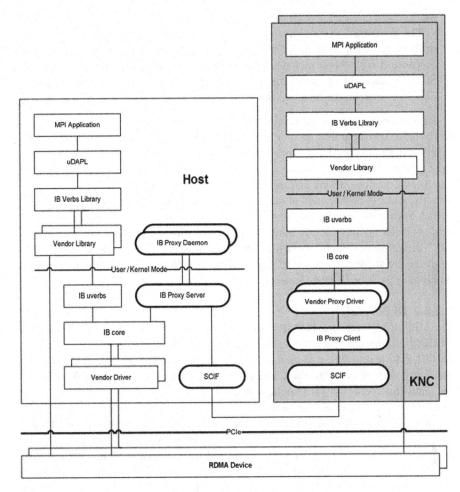

Fig. 2.23 MPI applications on MIC-Direct

monitors connections from the client side, relaying RDMA device insertion, removal, and event messages. The IB proxy server initializes core IB verbs to call the InfiniBand kernel layer representing the MIC IB proxy daemon and returns the results.

3. Vendor proxy drivers

A vendor proxy driver is the key module of MIC. Different vendor proxy drivers are used to support specific RDMA devices. Every vendor proxy driver registers RDMA device insertion, removal, and event messages of a specific PCI device through IB proxy daemon. Vendor proxy drivers use a programming interface provided by the IB proxy daemon to call IB verbs in kernel mode. Vendor proxy drivers transfer and translate all private data shared between the MIC coprocessor and the vendor drivers.

4. SCIF

The symmetric communication interface provides a sharing mechanism of nodes within the platform. A node can be a MIC device or a CPU devices system (because a CPU node includes the related mainboard, memory, power, etc.). SCIF abstracts the details of transferring through PCI-E, providing communication API between the host and the MIC device.

RDMA operations are achieved through a programming interface called verbs. Verbs can be separated into a privileged level and a non-privileged level. Privileged-level verbs are used to distribute and manage RDMA resources. Once the resources have been initialized, non-privileged verbs are used to finish the I/O operations. I/O operations can direct to or from the MIC user mode application directly, and can be executed at the same time with host I/O. Both of them can use kernel bypass and direct memory placement. The RDMA device provides the process isolation and address translation functionality required by I/O operations. MIC-Direct proxy verb operations between host and MIC coprocessors are achieved by treating every MIC coprocessor PCI-E card as another user mode process on the host IB kernel stacks.

2.2.8.2 OFED/SCIF

The MPI on MIC can use TCP/IP or OFED to communicate with other MPI nodes. OFED/SCIF drivers allow the InfiniBand host communications adapter (IBHCA) on PCI-E to access physical memory on the MIC coprocessor. When running on platforms without IBHCA, the OFED/SCIF driver can simulate IBHCA, making the platform able to run MPI applications.

OFED/SCIF provides software that simulates IBHCA to allow OFED applications to use high-performance communication capability based on RDMA (such as the InfiniBand). SCIF is a group of MIC communication API: a consistent and high-efficiency point-to-point communication between MIC coprocessor nodes. Thus, it is the communication path between the MIC card and the host side. Through the OFED layer on SCIF, many HPC applications based on OFED can easily use MIC architecture.

The OFED software stack is made up of many layers, including user layer applications and kernel driver libraries. Most of those layers are general codes shared by different vendors. Vendor-dependent codes are restricted in vendor hardware drivers and related user layer libraries (allow kernel bypass). Figure 2.24 shows the structure of the OFED/SCIF stack. Because the SCIF provides the same API for both the host and MIC side, this figure is suitable for both circumstances.

The bold, rounded rectangle in Fig. 2.24 shows detailed OFED/SCIF modules. Those modules include the IB-SCIF library, IB-SCIF driver, and SCIF (only includes kernel mode driver).

1. IB-SCIF library

The IB-SCIF library is the user layer library (required by the IB verbs library). It runs through the IB-SCIF driver. It defines a routine to allow the IB verbs library to call the IB verbs API defined on the user mode. It allows a specific vendor to optimize (including kernel bypass), making it work on the user layer.

Fig. 2.24 OFED/SCIF

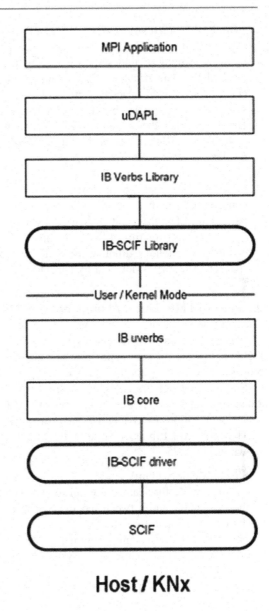

Host / KNx

However, the IB-SCIF library itself does not provide kernel bypass, and it uses the IB uverbs driver exposed interface, transmitting user mode requests to the kernel mode driver.

2. IB-SCIF driver

The IB-SCIF driver is the kernel module similar to a software-based RDMA device. In the initialization stage, it creates a link for every SCIF node pair and registers an "iWARP" device to the IB kernel driver. For some OFED operations (pure RDMA read/write), data is passed through the SCIF RMA functions. For other operations, data is passed as a data package. It creates a simple link between two SCIF nodes by identifying the communication context in the head of the data package. This is enough to support any number of logical connections. Under package protocol, small data can use scif_send() and scif_recv() to pass; large data use SCIF RMA functions to pass after signal exchange. When the two sides of a logical connection are on the same SCIF node (such as loopback), data is directly copied from the source to the objective position without SCIF transfer.

3. SCIF

The SCIF kernel module provides a communication interface among MIC coprocessors and between the MIC coprocessor and the host post. SCIF is not a part of OFED/SCIF, and it is used by OFED/SCIF only in communication channel nodes (in terms of SCIF, the host port is a node; each MIC coprocessor is a different node). Although the SCIF library provides a similar API in user space, that library is not used by OFED/SCIF.

2.2.8.3 Intel MPI Library Involving Intel MIC Architecture

The Intel MPI Library of Intel MIC Architecture only provides Hydra Process Manager (PM). Each node and coprocessor has a unique symbol or IP address. External (such as a command line) and internal (such as MPI_Comm_Spawn) processes, creating methods, and addressing capabilities can be executed on the nodes and coprocessors. It is the user's responsibility to distinguish the target architecture and execute their respective executable programs.

The applications can recognize their own platforms (CPU or MIC) at run time.

The Intel MPI Library of Intel MIC Architecture supports the architecture as shown in Fig. 2.25.

1. Shared Memory (shm)

This architecture can be used inside any coprocessor, among coprocessors of the same node, and between specific coprocessors and the host-side CPU where the coprocessor is located. The internal communication of the coprocessor is preset to call by a common mmap system based on shared memory. Other communications are called by a system similar to the SCIF scif_mmap(2) system. This architecture can be used separately or in combination with other architectures for a typical high-performance application.

The entire structure of SHM is shown in Fig. 2.26. Usually, the shared memory communication adds the SHM extension of SCIF, which supports multi-pass platforms, and each pass processor has a PCI-E interface. Not only can the SHM extension based on SCIF be used between the host processor and MIC coprocessor, but it can also be used among the connected coprocessors via the PCI-E bus.

2. DAPL/OFA

This architecture can be easily obtained from the two different interfaces of the Intel MPI library: the respective host channel adaptors (HCA) of the Direct Application Programming Library (DAPL, a high-performance remote direct

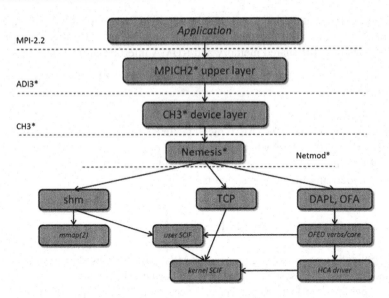

Fig. 2.25 The communication structure supported by the MPI library

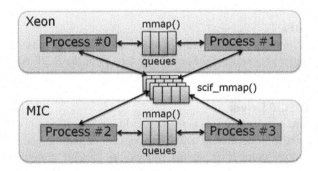

Fig. 2.26 Schematic of mmap

memory access interface library) and Open Fabrics Association (OFA), called the Open Fabrics Enterprise Distribution (OFED). In both cases, the typical remote memory access (RMA) protocol is mapped to a dedicated section in the system software layer, making a scif_writeto (2) and a scif_readfrom (2) system call.

3. TCP

This architecture is usually the slowest of all supported architectures; it works as a back-up communication channel when other high-performance architectures cannot work.

4. Hybrid Architecture

The above architectures can be reasonably combined to obtain better performance, such as shm:dapl, shm:OFA, and shm:tcp. This means using shared memory in the node and using another architecture (dapl, OFA, TCP, etc.) among the nodes. For more details, please refer to the Intel MPI documentation. It may also use the shared memory and RDMA interface at the same time.

Finally, the Intel MPI Library of Intel MIC supports the following types of I/O (input/output):

1. Standard file I/O: the standard file stream (including stdout, stdin, and stderr) is supported by Hydra PM. All typical features, such as redirection, work as expected. It is the same, whether on the MIC or the original CPU side.
2. MPI I/O: All MPI I/O features with MPI standards in all processes are the same as the underlying file system supports.

2.2.9 Application Programming Interfaces

Some application programming interfaces (APIs) have been used to assist a porting program to the MIC system. They are the sockets network interface and the message-passing interface (MPI). These reflect the industry standards and work in a variety of environments. Other internally developed APIs, such as SCIF, can be used by the public.

2.2.9.1 SCIF API
SCIF works as communication with the platform and has been exposed as low-level APIs for developer programming. For specific details of the SCIF API, please see Sect. 5.5.

2.2.9.2 OFED/SCIF Driver
MPI is a standard protocol for message-passing library interfaces; it works for two or more symmetric independent multi-processor computing nodes or among distributed memory computing agents (nodes) in parallel computing systems. MPI can be used for many HPI environments, providing high-performance scalability and portable communication methods in parallel applications.

MPI can use any one of the TCP/IP or OFED stacks to communicate with other MPI nodes in the MIC. The OFED/SCIF driver can be used on the IBHCA of the PCI-E to visit the physical memory of the MIC. When there is no IBHCA on the platform, OFED/SCIF can act as an analogue to an IBHCA, allowing MPI applications of MIC to communicate effectively on the platform.

2.2.9.3 NetDev Virtual Network
The virtual network driver provides a network stack to connect both sides of the PCI-E bus. The NetDev device driver analogues the hardware network driver providing the TCP/IP network stacks via the PCI-E bus. The sockets API and library provide the parallel applications among the computing nodes an end-to-end communication method based on industry standards. This set of API applies the TCP/IP protocol stack to provide applications an easy portability and scalability. This network stack also supports other standard network services.

The First MIC Example: Computing Π

3

In this chapter, we write and compile a MIC program to get an idea of how MIC programming works. We also introduce the MIC program running process to help give the reader a better understanding of the MIC program.

Chapter Objectives. Readers should gain a deeper understanding of:

- The MIC program framework and running process
- How to write a simple MIC program

From the Language B programming manual in 1972, "Hello World!" was the stepping-stone of learning a new computer language, and has since then become a symbol and has even achieved the status of a philosophy. Of course, when "Hello World!" became known throughout the programming world, a lot of people said that it was completely useless: even if you could write "Hello World", it does not mean that you knew how to use it. We will not discuss here who is right or wrong on this particular subject. We will not provide a traditional MIC "Hello World", because for parallel programing, it is not interesting to simply print out "Hello World" onto the screen.

Let us now look at the first MIC C/C++/Fortran example, in which we write a program to compute pi. Through this example, we will have an intuitive understanding of the programming style of the MIC program; grammatical and syntactical details are saved for later chapters.

E. Wang et al., *High-Performance Computing on the Intel® Xeon Phi™*,
DOI 10.1007/978-3-319-06486-4_3, © Springer International Publishing Switzerland 2014

//C/C++, Compile options: icpc main.cpp -openmp, run: ./a.out

[code]

```
1    #include <stdio.h>
2    #include <stdlib.h>
3    #include <math.h>
4    int main()
5    {
6        float pi = 0.0f;
7        int count = 10000;
8        int i;
9    #pragma offload target (mic)
10   #pragma omp parallel for reduction(+:pi)
11       for (i=0; i<count; i++)
12       {
13           float t = (float)((i+0.5f)/count);
14           pi += 4.0f/(1.0f+t*t);
15       }
16       pi /= count;
17
18       printf("Pi = %f\n", pi);
19   }
```

//Fortran,Compile options: ifort main.f90 -openmp, run: ./a.out

[code]

```
1    program mic_sample
2        REAL    :: t, pi = 0.0
3        INTEGER :: i, count =10000
4    !dec$ offload target(mic)
5    !$omp parallel do reduction(+:pi)
6        do i=0,count
7            t=(i+0.5)/count
8            pi=pi+4.0/(1.0+t*t)
9        enddo
10
11       pi=pi/count
12
13       print *, pi
14   end program mic_sample
```

This is surprisingly simple and important, because the key advantage of a MIC program is that it can make full use of MIC capabilities without changing too much the original CPU code.

As you can see, from the original sequential codes, only two directives were added (in bold, the ninth and tenth lines of the C code, and the fourth and fifth lines of the Fortran code) to make the program runnable on MIC. A closer look shows that the statement (the tenth line of the C code and the fifth line of the Fortran code) containing the "omp" directive is actually part of the OpenMP grammar. The statement with the "offload" command (the ninth line of the C code and the fourth line of the Fortran code) is the only real MIC statement. This command simply tells the compiler that the following code snippet will be running on MIC, and the rest will be passed to the compiler.

There is no difference in compiler options from the original CPU program. If MIC can support them, all of the compiler options on the CPU (if necessary) are loaded automatically onto MIC.

From the example codes above, we have learned:

1. The common MIC program running mode uses the "offload" statement (#pragma offload target(mic)/ !dec$ offload target(mic))
2. The work for the compiler is similar to that of the CPU, but for the "offload" statement, the code has two versions: the CPU and the MIC codes.
3. The parallelization of the MIC program is fulfilled by another method; the easiest method is to use it in conjunction with OpenMP.
4. The running processes of offload mode are:
   (a) The program is first run on the main function of the CPU.
   (b) The program reaches the "offload" statement, which informs it that the following code snippet will be run on MIC.
   (c) The driver program checks whether the MIC card exists. If yes, it calls the MIC code; if not, the CPU code is called.
   (d) Suppose the MIC card exists. If it is the first time a MIC program is being run, the driver program starts/wakes up the MIC card.
   (e) The MIC code is loaded onto the MIC card.
   (f) The memory on the MIC has opened more space, so the driver program copies the data from the memory on the CPU to the MIC.
   (g) The program on the CPU has been paused, and the program segment on the MIC starts to run.
   (h) After finishing the program segment on the MIC, data is copied from the memory on the MIC to the CPU.
   (i) The MIC card returns to low-power consumption status, and the program on the CPU resumes.
   (j) Program ends.

Thus, behind the simple "offload" statement, there is a huge workload. The MIC program hides a lot of trivial details so the programmer can focus on the algorithms and code, which improves development efficiency.

All in all, because the compiler hides a lot of details, it makes the MIC program, whether it is the code or the compiling process, very similar to the current CPU program, especially in parallel programming. Thus, the learning curve of MIC is very smooth, making it extremely easy for programmers who have previous experience in developing parallel programs on CPUs to easily learn and use the MIC.

Fundamentals of OpenMP and MPI Programming

4

This chapter outlines the basic usage of OpenMP and MPI, including a brief introduction and grammar. It will help readers to learn or review basic OpenMP and MPI knowledge and lay down a solid basis for compiling a MIC program.

Chapter Objectives. After reading, you should have a deeper understanding of:

- the basic knowledge of OpenMP
- the basic use of OpenMP, and how to compile a simple OpenMP program
- the basic knowledge of MPI
- the basic use of MPI, and how to compile a simple MPI program

4.1 OpenMP Foundation

Why is there a special chapter for OpenMP in a MIC book? Because OpenMP is a powerful tool for parallelization on a MIC program and a good method for transforming a sequential program into the corresponding MIC version. Of course, the operating system on the MIC card also belongs to the Linux class (a custom embedded Linux system), so MIC can also support pThread and other parallel programming tools. Yet, leaving MIC aside, when facing highly parallel tasks, no matter the programming simplicity or the mode of thinking, using OpenMP is always preferable.

So, what is the relationship between OpenMP and MIC programming? MIC extending C and Fortran, strictly speaking, is for instructing some code segment that is running on the MIC card, as well as for data transmission. OpenMP is the part of the code that uses the core of the MIC card for parallel execution.

E. Wang et al., *High-Performance Computing on the Intel® Xeon Phi™*,
DOI 10.1007/978-3-319-06486-4_4, © Springer International Publishing Switzerland 2014

We note here this is not an OpenMP introduction manual, so this book only contains a basic explanation of how OpenMP works. If you need more advanced techniques, you should refer to related manuals or books on this topic

4.1.1   A Brief Introduction to OpenMP

OpenMP is an API (or more precisely, OpenMP is an extension of language) for writing portable multi-threaded applications. It provides a simple method so the programmer doesn't need to worry about a complicated thread creation, synchronization, load balance, and destruction process. In the beginning, OpenMP was based on a Fortran standard; now, after development, it can support C and C++ too. The latest version of OpenMP is 3.1, which supports Fortran, C, and C++. The OpenMP program design module provides a set of platform-independent Pragmas (which work for C and C++), Directives (work for Fortran) (Note: Pragma and Directive hereinafter will be referred to as quotation), Function Calls, and Environment Variables. It instructs the compiler on how and when to use parallelism in running applications.

4.1.2   OpenMP Programming Module

Based on shared memory and threads, OpenMP is a parallel programming module that supports signal instruction stream. OpenMP is based on the fork/join module, which means that parallel tasks create parallel threads at the beginning, execute the parallel tasks, and then terminate those threads when the task is completed. The task is finished if all the threads have been terminated. OpenMP requires the programmer to explicitly control the parallel codes: the parallel mode can be anticipated, but the programmer controls the data dependency.

4.1.3   Brief Introduction to OpenMP Grammar

4.1.3.1 Overview of OpenMP Program

OpenMP uses quotation to parallelize programs. The most common way is to use loop parallelization (e.g., the For loop in C/C++, the Do loop in Fortran).

For example:

C/C++:

[code snippet]

```
1 #pragma omp parallel for
2 for(i=0;i<LEN;++i)
3 {
4     a[i]=b[i]+c[i];
5 }
```

Fortran

[code snippet]:

```
1 !$OMP parallel do
2 do i= 1 , LEN
3     a(i)=b(i)+c(i)
4 end do
5 !$OMP end parallel do
```

The above code is for OpenMP marked loop parallelization; that is, more "a [i]=b[i]+c[i]" parallel execution. The specific parallel mode depends on the threads and OpenMP options. In the best circumstances, all of the loop statements will be executed at the same time.

OpenMP's limitations for the loop are:

1. The loop variable must be an integral type.
2. The comparison operation of the loop statement (the second operation of the For statement) must be a loop variable: $>$, $>=$, $<$, or $<=$.
3. The loop step must be an addition or reduction integer, and its value must be a loop invariant.
4. If the comparison operation is $<$ or $<=$, the value of the loop variant must be increased each iteration, and vice versa.
5. The loop must be single-entry and single-exit, which means the loop cannot quit halfway.

Although it looks like there are a lot of limitations, most loops can meet these conditions, or match the OpenMP parallel conditions after some revision.

The loop that matches the above conditions can use OpenMP to parallel, but if you want to have a better parallel effect with OpenMP, you should consider how suitable parallelization is for the loop.

There is an easy way to identify whether the loop is suitable for parallelization. To analyze the parallelization of loops, check whether the loop body (not the For or Do statement) follows a sequence, when the values of the loop variables are different: if one loop must run before another loop, or if the two loops use the same memory space. If the loop follows a sequence, it is called "data dependent" (e.g., one loop uses a[i] and a[i+n]). If we cannot solve this date dependent problem, the parallelization will not be as good, or it cannot be paralleled at all. (Again, OpenMP only provides the parallel function; it is unable to make the decision of whether the program can be paralleled. And even if the program can be paralleled using OpenMP, the result may not be correct.) If the loop body uses the same memory space, it is called "data competition". This can usually be eliminated by declaring private variables.

4.1.3.2 Basic Grammar of OpenMP

We just had an intuitive impression of OpenMP programming by means of a small example in the previous section, and now we will explain the basic grammar of OpenMP below.

The basic usages of OpenMP are:

```
C/C++:
[code snippet]
#pragma omp parallel
{

    //Code segment

}
```

```
Fortran:
[code snippet]
!$OMP PARALLEL
    !Code segment
!$OMP END PARALLEL
```

The region marked by the parallel command is the parallel region. We could add private, firstprivate, lastprivate, and num_threads after the parallel statement to control data and parallel region properties.

Private(list): its attribute is the parameters list, separated with commas, and it represents the variables in the parallel region that are private. Each thread has a copy.

Firstprivate(list): has a similar attribute as private. The initial value of the corresponding variable stays the same before entering the parallel region.

Lastprivate(list): has a similar attribute as private, but when the parallel region ends, the corresponding variable keeps the number of last ended thread. Please note: the last ended thread is not always numbered the last.

Num_threads(num): set up to open the num threads in the parallel region.

Schedule(flag): set up thread distribution methods (which loop will be assigned to run on any CPU core). The options are: dynamic[,n], static[,n], guided[,n], and runtime. $n=1$ if there is no special indication.

- Dynamic: dynamic assignment. It follows the iteration order, one task is N times of the iteration, and assigns tasks to threads using a principle based a first-come-first-served. Except for the first time, it is not sure which loop should be executed in which thread.
- Static: static assignment. It follows the iteration order and one task is N times of iteration. If each thread has m tasks to run on average, then the first m tasks are assigned to the first thread, the $m+1$-$2m$ tasks to the second thread, and so on. The last thread may run less than m tasks. This method will lead to a load unbalance in the "unbalanced tasks" situation, but if the tasks are averaged, it will reduce the cost of task assignment.
- Guided: guided assignment with dynamic partitioning. Assume a start of m threads: the first thread is assigned to the previous $1/m$ of the whole iteration, the second thread to the $1/m$ of the rest iteration, and so on. The minimum assignment unit is n times iteration. The task assignments are determined on a first-come-first-served basis. This method not only reduces the number of task assignment occurrences, but also does so dynamically: it strikes a balance between the dynamic and static portions.
- Runtime: not an assignment model, but it specifies using one of the above ways when working with environment variables. We can use the following assignment when running a program: setenv OMP_SCHEDULE "dynamic, 2". In the parallel region, the most common clause is omp for/omp do, as shown below:

```
C/C++:
[code snippet]
#pragma omp parallel
{
    #pragma omp for
    for()
    {
        ...
    }
}
```

```
Fortran:
[code snippet]
!$OMP PARALLEL
  !$OMP DO
  do
  ...
  end do
  !$OMP END DO
!$OMP END PARALLEL
```

This usage is more common, so it is often merged with the first example (note: there are no other parameters between parallel and do/for). This book introduces this usage because it is the most commonly used method for MIC. But MIC can support other usages that meet OpenMP requirements.

The format requires two instructions. First, the merged writing and not-merged writing parts are not entirely equivalent: if we use not-merged writing, we can have multiple parallel regions in the parallel zone, but merged writing can only have one. Second, a separate writing of omp for/opm do or merged writing under the declaration line (i.e., the second line, comment not included) must be a for/do loop; otherwise, it is a syntax error.

Besides private, firstprivate, and lastprivate, the parameter after the for/do statement can be a reduction statement. It can be used just like the reduction (+:sum): the beginning is the reduction symbol, followed by the reduction of the variable. The supported symbols are +, -, *, &, |, ^, &&, and ||. OpenMP maintains a copy of each thread and initializes according to the operator, and runs the reduction operations at the end of the parallel region. Each parameter is separated by a space.

4.1.3.3 Library Functions of OpenMP

Besides compiler instructions, OpenMP also provides a set of library functions and environment variables. The most commonly used library functions are:

- int omp_get_num_threads(void)
  Return to the current number of used threads.
- int omp_set_num_threads(int NumThreads)
  Before entering the parallel region, set the number of threads that will be used in the parallel region. Its priority is higher than that of the environment variables.
- int omp_get_thread_num(void)
  Return to the current thread number.
- int omp_get_num_procs(void)
  Return to the number of available processor cores (logical thread, considers double if opening the hyper-thread).

4.1.3.4 The Environment Variables of OpenMP

The Environment Variables:

- OMP_SCHEDULE
  Controls the direction of loop scheduling
- OPM_NUM_THREADS
  Sets up the number of threads in the parallel region
- KMP_AFFINTY

Sets up the binding patterns between OpenMP threads and core computing. The acceptable values are:

none: not binding

scatter: scatter pattern

compact: compact pattern

balanced: balanced pattern (only on MIC; CPU does not support this, and this may be ignored)

There are several ways to execute KMP_AFFINITY:

1. Compact mode: As Fig. 4.1 shows, in compact mode a thread tries to fully use the cores; it uses up one core and then moves on to the next one. This can easily cause load unbalance. However, if there are data exchanges or data sharing between neighboring threads, these will benefit due to the presence of the threads being on the same core.
2. Scatter mode: As Fig. 4.2 shows, in scatter mode threads are put into the least-used cores for the best load balance. This is suitable for threads with no dependency on their neighbors.
3. Balanced mode: As Fig. 4.3 shows, balance mode is similar to scatter mode in trying to take into account both load balancing and data localization. It performs well in most applications.

Both OMP_SCHEDULE and KMP_AFFINTY try to optimize load balancing. However, there are many differences besides the programming method: OMP_SCHEDULE focuses on how loops are distributed among threads, while KMP_AFFINTY focuses on how threads are distributed among cores. If SCHEDULE is the logical distribution, then AFFINITY is the physical distribution. They cooperate with each other, and the best load balancing can be achieved only if both are set up properly.

4.2 Message-Passing Interface Basics

Message-passing interface (MPI) is a popular processes communication API in parallel computing. MPI can set up communication between different nodes through a network. This is the basis of distributed computing. The programming models here can be in shared or unshared memory mode. MPI is also the most popular communication library on MIC, so we will give a brief introduction.

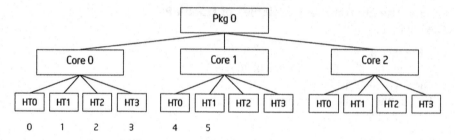

Fig. 4.1 Compact mode

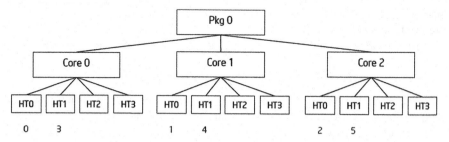

Fig. 4.2 Scatter mode

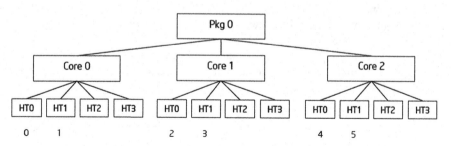

Fig. 4.3 Balanced mode

The MPI standard defines a group of kernel functions' grammar and definitions. It is very helpful in writing message-passing programs. There are more than 125 functions in the MPI library, but the number of key functions is much lower. Actually, a fully functional message-passing program can be written with only the six functions in Table 4.1. These functions are used to initialize and terminate the MPI library, get information from the parallel computing environment, and send and receive messages.

Table 4.1 MPI functions

Name	Functionality
MPI_Init	Initialize MPI
MPI_Finalize	Terminate MPI
MPI_Comm_size	Get number of processes
MPI_Comm_rank	Get process ID
MPI_Send	Send a message
MPI_Recv	Receive a message

In this section, we introduce these six functions; please refer to an MPI manual for other functions.

4.2.1   Start and End MPI Library

Before calling any other MPI functions, MPI_init needs to be called first to initialize the MPI environment. Calling MPI_Init more than once in a program yields an error. After the computing is finished, call MPI_Finalize to run the ending task and terminate the MPI environment. There should not be any MPI calls after MPI_Finalize, including MPI_Init. MPI_Init and MPI_Finalize must be called by all processes, or the state of MPI is uncertain. The declarations of those two functions in C is the following:

```
int MPI_Init(int   *argc, char   ***argv)
int MPI_Finalize()
```

After MPI_Init or MPI_Finalize is executed successfully, either MPI_SUCCESS or an error code will be returned. The specific error code depends on the version of the MPI library.

The parameters argc and argv of MPI_Init are the command line parameters of the C programming language. When MPI runs, it removes all command parameters that the MPI library can handle from argv, and reduces the number in argc accordingly. Therefore, the programmer must deal with command line parameters after calling MPI_Init.

Error declarations in Fortran:

```
MPI_Init(IERROR)
MPI_Finalize(IERROR)
```

The only parameter is the error code; MPI_Init will not handle command line parameters.

From the above two functions we can see that MPI has a naming and parameter standard. All MPI functions, data types, and constants start with "MPI_". If run successfully, MPI_SUCCESS will be returned. In C, the header file "mpi.h" needs to be included; in Fortran, the header file "mpif.h" needs to be included.

4.2.2   Getting Information About the Environment

Before explaining MPI_Comm_size and MPI_Comm_rank, we must understand the idea of the communication domain. A communication domain is a group of processes that can communicate with each other. Information related to the communication domain is stored in MPI_Comm type variables; variables of this type are called "communicators". These communicators are used as parameters for all MPI functions in message passing and uniquely identify the processes that participate in the communication operations. Every process can belong to a different communicator, or they can overlap. A simple analogy would be that a communicator is like a school phonebook: a number can belong to an entire family and can also belong to a classmate.

Usually, all the processes require communication. Therefore, MPI defines the default communicator MPI_COMM_WORLD, which includes all MPI processes. If the programmer needs to send a message within some of these processes and filter out the others, a customized communicator needs to be defined.

MPI_Comm_size returns the number of processes, and MPI_Comm_rank returns the process IDs. Their declarations in C are the following:

```
int  MPI_Comm_size(MPI_Comm  comm,  int  *size)

int  MPI_Comm_rank(MPI_Comm  comm,  int  *rank)
```

The Fortran calls are similar: just add a return code in the parameters.

MPI_Comm_size returns the number of processes in the communicator "comm". If the communicator is MPI_COMM_WORLD, this call returns the number of processes launched by the MPI.

MPI_Comm_rand gets the rank in the communicator "comm". The range of rank numbers is zero to one less than the communicator size. Normally, the rank-zero process is treated as the main process, but the main process can only be determined after getting the rank numbers using this function.

4.2.3   Send and Receive Messages

The basic function of sending and receiving messages are MPI_Send and MPI_Recv. The declaration of those two functions is shown in the following:

```
int  MPI_Send(void  *buf,  int  count,  MPI_Datatype  datatype, int  dest,  int  tag,
              MPI_Comm  comm)
int  MPI_Recv(void  *buf,  int  count,  MPI_Datatype  datatype,  int  source,  int  tag,  MPI_Comm
              comm, MPI_Status  *status)
```

MPI_Send sends the data stored in the buffer appointed by "buf". The data is stored continuously in the buffer with the datatype defined by "datatype", and it can be treated as an array of that datatype. The number of elements in the buffer is given by "count". However, "count" can be set as zero; in this case, an empty message is sent. The possible choices for datatype are shown in Table 4.2.

Table 4.2 MPI datatypes corresponding to C

MPI datatype	C data type
MPI_CHAR	signed char
MPI_SHORT	signed short int
MPI_INT	signed int
MPI_LONG	signed long in
MPI_UNSIGNED_CHAR	unsigned char
MPI_UNSIGNED_SHORT	unsigned short int
MPI_UNSIGNED	unsigned int
MPI_UNSIGNED_LONG	unsigned long int
MPI_FLOAT	float
MPI_DOUBLE	double
MPI_LONG_DOUBLE	long double
MPI_BYTE	–
MPI_PACKED	–

Nearly all MPI datatypes have corresponding C datatypes, but MPI_BYTE and MPI_PACKED have no definition in C.

MPI_BYTE corresponds to a byte (8-bit), while MPI_PACKED corresponds to a set of data. MPI_BYTE is not difficult to understand. Consider the universality of different platforms: using MPI_BYTE can guarantee the data of one byte size. However, if we use MPI_CHAR, it may use two bytes in some platforms, thus affecting portability. MPI_PACKED has a similar but not exactly equivalent idea of "struct" in C. MPI_PACKED data uses MPI_PACK to pack and MPI_UNPACK to unpack; the data inside the package may not continue.

MPI datatypes corresponding to Fortan are shown in Table 4.3.

Table 4.3 MPI datatypes corresponding to Fortran

MPI datatype	Fortran data type
MPI_INTEGER	INTEGER
MPI_REAL	REAL
MPI_DOUBLE_PRECISION	DOUBLE PRECISION
MPI_COMPLEX	COMPLEX
MPI_LOGICAL	LOGICAL
MPI_CHARACTER	CHARACTER(1)
MPI_BYTE	
MPI_PACKED	

The destination of the message sent by MPI_Send is defined by "dest" and "comm". The value of dest is based on the rank within the communicator comm. The value of "tag" is the ID of the message; every message has an integer number tag, which can be used to separate different messages. The range of tag is from zero to MPI constant MPI_TAG_UB. Although the value of MPI_TAG_UB may vary, the minimum is 32767. The value of tag is set by the programmer.

MPI_Recv receives the message sent by a process; this process is defined with the rank, by source, and the communicator, by comm. The message ID is defined by tag; if more than one message from the same process has the same tag, a random one will be received. MPI allows source and tag to use a wild card: if source is set to MPI_ANY_SOURCE, any process of the communicator can be the source. Similarly, if tag is set to MPI_ANY_TAG, a message with any ID will be received. The received message is stored in the buffer appointed by buf, and the buffer size is defined by number of elements in "count" and the datatype "datatype" (the buffer has to be allocated by the programmer). However, this is the size of the buffer and not the size of data to be received; the data to be received should be less than or equal to this size, which does not need to be known prior to calling the receive function. If the message to be received is longer than the buffer size, it will return MPI_ERR_TRUNCATE, showing an overflow.

After the message is received, we can use variable status to get the operation information of MPI_Recv. In C, status is defined as MPI_Status, structured like so:

```
typedef struct MPI_Status{
  int MPI_SOURCE;
  int MPI_TAG;
  int MPI_ERROR;
};
```

MPI_SOURCE and MPI_TAG keep the source and ID of the message. They are especially useful when source is set to MPI_ANY_SOURCE or tag is set to MPI_ANY_TAG. MPI_ERROR keeps the error code of the received data.

The status parameter can also return the length of the message, but not from the status variable directly. It requires a call to MPI_Get_Count, defined as the following:

```
int MPI_Get_Count(MPI_Status  *status,   MPI_Datatype datatype, int  *count)
```

The three parameters of MPI_Get_Count:

- Status: the status returned by MPI_Recv.
- Datatype: the datatype of the message.
- Count: returns the actual number of elements received.

Programming the MIC

5

Now that we have reviewed the basics of OpenMP and MPI, we can start to understand the MIC programming models and application modes. Once we learn the basic syntax of MIC and the related MPI usages on MIC, we can readily write a MIC program. Finally, we will introduce the API–SCIF system programming on MIC for those readers who have special p needs for high-performance computing.

Chapter Objectives. After reading this chapter, you should have a deeper understanding of:

- MIC programming models and application modes
- how to write, compile, and run the MIC programs
- how to use MPI programming techniques on MIC
- SCIF programming

5.1   MIC Programming Models

Just like most computing systems, MIC architecture programming models can be divided into two types: application programming and system programming.
1. Application Programming
    Application programming uses Intel Composer XE 2013 or third-party development tools to develop user applications or codes. Application programming is the most used programming model for the programmers, and is also the focus of this book.
2. System programming
    System programming mainly introduces "how to use MIC architecture". System programming uses a low-level API (such as SCIF) and is included in the Intel MIC Platform Software Stack (MPSS). For more details, please see the relevant sections on SCIF.

E. Wang et al., *High-Performance Computing on the Intel® Xeon Phi™*,
DOI 10.1007/978-3-319-06486-4_5, © Springer International Publishing Switzerland 2014

5.2   Application Modes

MIC has extremely flexible programming modes. The MIC card can be used as a coprocessor or as a separate node. The most common MIC application mode considers the MIC card as a coprocessor. Based on the program's instructions, the CPU will run parts of the code on the MIC side, so there are two kinds of equipment at the same time: the host (CPU side) and the device (MIC side). The MIC card not only can act as a coprocessor, but its X36 instructions architecture—which has the same function as the CPU and μOS onboard operating system based on Linux—gives the MIC card more powerful functions. Therefore, the relationship between them can be summarized as one of the five shown in Fig. 5.1.

Application modes can be simply divided into three modes: CPU-centric, MIC-centric, and CPU and MIC peer-to-peer.

1. CPU-centric mode

This mode is divided into two circumstances: pure CPU computing, which usually applies to serial and parallel computing programs, and its parallel part has a low degree of parallelism, and CPU-centric computing with MIC cooperation, which applies to serial computing programs containing highly parallel computing segments and has a high degree of parallelism. The CPU starts the main function of the program. When it reaches the highly parallel computing part, the parallel computing runs on the MIC card.

2. MIC-centric mode

Pure MIC Computing This applies to highly parallel computing programs, and these programs run directly on the MIC. If you pass part of the computing to the CPU, it creates additional costs, such as increasing the communication between the CPU and MIC, or waiting for computing synchronization between the CPU and MIC.

MIC-centric Computing with CPU Cooperation This applies to highly parallel computing programs with a partial serial computing segment. The program's main function is initiated on the MIC, and passes to the CPU when it reaches the serial computing part; the CPU acts as a coprocessor.

3. CPU and MIC peer-to-peer mode

This applies to multiple parallel computing programs, such as the MPI program. The program's main function is initiated by the CPU and MIC at the same time.

In summary, with a MIC coprocessor the program has five different operation modes: CPU-native mode, CPU primary and MIC supplementary mode, CPU and MIC peer-to-peer mode, MIC primary and CPU supplementary mode, and the MIC in native mode.

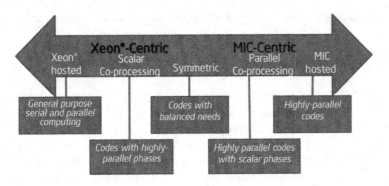

Fig. 5.1 MIC application modes

Simply put, the main control terminal can be on the host (CPU side), as well as the device (MIC side); the calculation function can be assigned to any of them. When the task is initiated from the device, we can even say that the MIC side is the host while the CPU side is the device. However, in order to avoid confusion, we will always consider the CPU side as the host and the MIC side as the device.

Ideally, the CPU and MIC work together. The CPU is responsible for the strong logical transactions and serial computing, and the MIC is responsible for performing the parallel tasks. The CPU and MIC have their own separate memory address spaces: the host memory and the device memory. Data is exchanged between the host and device through an API function or compiler instructions. Memory in the device can be used almost in the same way as with the host (including Linux system call); existing codes and programming practices do not often need to be changed.

The operation mode of the code segment on MIC is not much different than when it is on a multi-core CPU. Even from a programming perspective, it is exactly the same as a multiple thread program (no matter the number of actual computing cores). Therefore, we can say that there is no essential difference between the MIC programming model and the CPU programming model, which we are already familiar with. What the MIC program adds (with the CPU and MIC working together) is an indication that the code is running on the device, and the corresponding data transmission statement. Even these data transmission statements are just instructions and not explicit transmissions in the traditional sense.

With the multiple instruction stream and multiple data stream (MIMD) architecture, the MIC device can use SIMD parallel mode, pThread, MPI, or other modes to run different instructions on different computing cores.

The MIC coprocessor has the same $\times 86$ instruction set as the CPU, although there is a slight difference in assembly instructions (mainly in the vectorization instruction part). But in general, there is no essential difference. Therefore, the MIC is not replacing the CPU, but instead is used to compute together with the CPU; it is an expansion of the CPU application mode.

The following sections explain the five operating modes in greater detail.

5.2.1   CPU in Native Mode

In the CPU-native mode programs only run on the CPU side. We used this mode until the MIC was developed. In this mode, although the hardware provides the resources for an MIC card, the program does not use the MIC card. This mode always uses the CPU to compute, so it can be used for only serial and a small amount of parallel computing applications; we will not explain in detail. When compiling with CPU-native mode, one needs to include the compiler option "-no-offload".

5.2.2   CPU Primary, MIC Secondary Mode

The mode when the CPU is primary and the MIC is secondary is the most common one employed in MIC programming. It applies to serial programs containing a partial highly parallel program. The MIC card is the coprocessor that assists and helps the CPU. This mode is often called the offload mode, but this model does not simply use the offload statement. The programming form of this mode is similar to OpenMP: it compiles directive statements and marks the code segment to be run on the device. Typically, the offload statement combines with the OpenMP statement by using the MIC device to perform the parallelization.

Offload: a statement derived from the CPU+MIC program by using the device to perform the marked code segment containing the "offload target". Thus, "offload" can mean several things:

1. Indicates the offload code.
2. Indicates the mode that is programming and running the offload statement, namely the coprocessor utilization mode.
3. Indicates the action of putting and running the code segment on the coprocessor.

Depending on the context, this book uses the term offload according to one of the above meanings.

Offload mode is usually used in the single-node structure and/or when the program has been modified by OpenMP. Offload mode works on single machine–single MIC-card and single machine–multiple MIC-cards, but it is not used to control the multi-node with offload statements. Therefore, if it is used on a multi-node parallel application, it needs to transfer data in the peripheral framework (e.g., MPI) and use the offload + OpenMP mode internally (e.g., the MPI process).

If you want to transplant the offload mode of the existing serial program to the MIC program, you will need to confirm the part to be parallelized first, and then rewrite the serial program to an OpenMP parallel program. After confirming the input and output variables, write the corresponding offload statement, and then the parallel code segment can be easily transplanted to the MIC. Of course, the MIC not only supports OpenMP, but also pThread, Cilk library, etc., to aid programmers in parallelization, according to what they're already used to.

The offload model is based on the CPU side, considering the MIC side as a coprocessor. The CPU side is the control side, completing control tasks and data transmission tasks (and perhaps also some computing tasks). The MIC end is the main computation side, completing the parallel computing tasks given by the CPU.

This mode is very similar to the existing GPGPU's operation mode. Once there is confirmation of the part of the program that needs to be parallelized (usually a loop), the computation work can be passed to the MIC. The basic flow is in the serial program; when performing the part that needs to be parallelized (marked by the keyword offload), the code and corresponding data are sent to the device to be parallelized; after its return (it may have an asynchronous call), the main thread continues to perform.

5.2.3 CPU and MIC "Peer-to-Peer" Mode

The CPU and MIC "peer-to-peer" mode, or the peer-to-peer mode, means that the CPU and MIC perform the same tasks; it considers the MIC device as a node equivalent with the CPU. The MIC device, which benefits from the μOS, can also run a program similar to a CPU-side program, and the MIC device has its own IP address, so this peer-to-peer mode is possible.

This mode operates in a similar fashion to a cross-nodes parallel program. It uses the MPI function interface to run data communication, and an offload statement is used to copy all of the programs to be performed on the MIC side. To see the compile options and method of program deployment, please consult Sect. 5.2.5. Program operation mode is the same as that of the MPI multi-node program.

5.2.4 MIC Primary, CPU Secondary Mode

The mode when the MIC is primary and the CPU is secondary is the opposite of the mode discussed in Sect. 5.2.2, where the CPU is primary and the MIC is secondary. In this operation mode, the MIC side acts as the host, and the CPU is considered a coprocessor. For programmers, this mode may be more difficult to understand than the peer-to-peer mode. However, if there is an application for this mode, it is parallel computing, with only a few serial code segments (e.g., File read–write) running on the CPU side. Thus, we would consider the CPU as a coprocessor and the MIC side as a host.

This mode also reflects what is special about the MIC, which makes it possible for many computing devices (including the CPU, the MIC, etc.) to be considered as computing resources. There is no distinction between major and minor (at least, computation-wise; the node is unable to start without the CPU), and the mode uses the most suitable device to compute.

5.2.5   MIC-Native Mode

As we known, the MIC card has its own operating system based on Linux, as well as its own IP address. So MIC supports on-card operation mode for transferring programs and data onto the MIC card manually, and then running the programs directly on the MIC card.

There are many advantages to running directly on the card. It saves time normally used for data transmission; it is suitable for processing the same data multiple times, it can set the device as the control side and launch the computing task, and it can set the CPU side as a coprocessor (the MIC primary, CPU secondary mode). It can be used in combination with SCIF (like the socket network programming interface) or MPI, considering the CPU and MIC device in a node as a small-sized computing network and then applied on Server/Client mode. All the programs can also be run entirely on the MIC device (MIC-native mode). Running the MIC program natively on the card is usually the best choice if the program's whole algorithm is parallel, or if the part of the code that could be run on the CPU will cost too much for transmission and synchronization.

For this mode, the compiler option—mmic must be added: this compiles the program to be run only on the MIC side and not the CPU side. Therefore, when running programs with MPI, two codes need to be compiled; the CPU side and the MIC side will run the corresponding codes while the MPI program starts in order to run the programs correctly.

The SCP command can be used to transfer programs and data to the MIC card. It is important to note that besides the user's own programs and data, the corresponding shared library usually needs to be transferred as well. Examples are libiomp5.so and some default MIC libraries under the compiler installation path lib/mic.

5.2.5.1 MIC-Native Program Example: Computing π
In this section, we demonstrate the entire operation process of the MIC native mode. Its usage also works for the peer-to-peer mode, the "MIC is primary" mode, and any other modes in which the programs need to start on the MIC side.

The following is the program source code, written in C, for the example of computing pi given in Chap. 3. Please note that we use the OpenMP version here, and not the MIC version with the offloaded statements (i.e., delete the ninth line of the code).

```
     [code]
1    #include <stdio.h>
2    #include <stdlib.h>
3    #include <math.h>
4    int main()
5    {
6         float pi = 0.0f;
7         int count = 10000;
8         int i;
9    #pragma offload target (mic)
10   #pragma omp parallel for reduction(+:pi)
11        for (i=0; i<count; i++)
12        {
13             float t = (float)((i+0.5f)/count);
14             pi += 4.0f/(1.0f+t*t);
15        }
16        pi /= count;
17
18        printf("Pi = %f\n", pi);
19   }
```

Then compile with:

```
#icc pi.c -openmp -mmic -o mic_pi
```

Please note that a "-mmic" option has been added here. If the CPU side runs this code, it will lead to an error:

```
# ./mic_pi
-bash: ./mic_pi: cannot execute binary file
```

This file can only be run on the MIC. The MIC card's default IP address is 172.31.1.1; for more details on setting the IP address, please see Sect. 5.4.3.

Upload with SCP command:

```
#scp   ./mic_pi   172.31.1.1:/tmp
```

Generally, the uploaded directory is under the /TMP path (i.e., everyone has permission to compile) or your home path. At this point, you can use the sSH remote execution command with the -x parameter (lowercase), but the program may report an error.

```
# ssh –x 172.31.1.1   /tmp/mic_pi
/tmp/mic_pi: error while loading shared libraries: libiomp5.so: cannot open shared object file:
No such file or directory
```

Because there is no corresponding OpenMP library on the MIC card, we need to upload the MIC version of the OpenMP library from the CPU side to the MIC card (please pay attention to the path; don't upload the CPU version of the library).

```
#scp /opt/intel/composer_xe/compiler/lib/mic/libiomp5.so   172.31.1.1:/lib64
```

Please note that the installation path on the CPU side may be different. Running the program once more should now give the correct result.

```
# ssh –x   172.31.1.1   /tmp/mic_pi
Pi = 3.141592
```

Of course, you can also log onto the MIC card with SSH before running the programs.

5.3 Basic Syntax of MIC

As mentioned above, MIC does not have a separate programming language. MIC programming is an extension of the C/C++Fortran language, adding compiler-directed or direction commands. This is the simplest way to make an existing program utilize MIC computing resources without changing too much. It is very similar to the OpenMP mentioned before. Of course, MPSS (Intel MIC Platform Software Stack) also offers some advanced API function interfaces (including some assembler instructions) so that users of different levels can make targeted and more complicated performance optimizations, depending on their application needs. It is also similar to OpenMP, so learning just one or two syntaxes should cover most needs, but to meet personalized demands, additional reading of other materials may be helpul.

5.3.1 Offload

Offload is the most basic keyword in MIC language extensions: it is essential to the MIC program, representing the program code in offload scope (the first code segment followed by the nearest offload statement) that must be run on the MIC card. The offload statement works on the CPU and the MIC master–slave mode. The most common application mode sets the CPU as a control side and the MIC as a coprocessor. This mode covers more than 90% of MIC, so we will introduce the related syntax more precisely in this chapter.

5.3.1.1 Offload Statement

Below are the basic usages:

```
C/C++:
[code snippet]
#pragma offload
```

```
Fortran:
[code snippet]
!dec$ OFFLOAD
Or:
! DIR$ OFFLOAD BEGIN ......
! other statements except OpenMP statements, eg. DO, CALL
!DIR$ END OFFLOAD
```

There is a slight difference with Fortran syntax: in the first case, the offload statement must be followed immediately by OpenMP statements or function call statements. In the second case, OpenMP statements cannot be used.

This chapter starts with a simple example to explain each MIC syntax.

First, let's see a piece of code without MIC instructions:

```
//C/C++
[code]
//demo1
1    #include <stdlib.h>
2    #include <stdio.h>
3    #define LEN 5
4    int main(int argc , char* argv[])
5    {
6        int i;
7        float x=2;
8        float arr[LEN];
9        for(i=0;i<LEN;++i)
10       {
11           arr[i]=i*3.0f/2.0f;
12       }
13       return 0;
14   }
```

```
!Fortran
[code]
! Note: The subscript of Fortran array starts from 0, so the C programs and Fortran programs in the examples
of this book are not exactly equivalent; we will not repeat them again.
1          program demo1
2          integer,parameter ::len=5
3          integer :: i
4          REAL, DIMENSION(len)::arr
5          do i=1,len
6             arr(i)=i*3.0/2.0
7          enddo
8          end program
```

The code is very simple and is mainly in a For loop (a Do loop in Fortran); in a parallel computing optimization process, the vast majority of existing serial code is in this form of loop (although it is much more complicated). This is a typical SIMD model: each loop iteration has the same instructions but with different data. Because the data are independent, this program has the best parallelism. Of course, the real-world program is not so perfect, and there is usually a lot of branching and data dependency, so the simultaneous effect is not as good as it is in theory. The first version of the program is mainly to verify if the program can be performed correctly, so the loop body is replaced with the print language for easy verification.
That is:

```
[code snippet]
for(i=0;i<LEN;++i)
{
    printf("Index:%d\n",i);
}
```

```
[code snippet]
do i=1,len
        print *,"Index:",i
    enddo
```

Compile this program.

```
icc –o demo1 demo.c
```

Or:

```
ifort –o demo1 demo.F
```

Run as follows: (we will not describe the compiler operation in detail in the future; please always refer to these steps or write the Makefile file.)

```
./demo1
```

Not surprisingly, we see that the output is (0–4 in the C program):

```
Index:1
Index:2
Index:3
Index:4
Index:5
```

We are going to put this piece of program to the MIC. Because this program is very simple, data transmission is not necessary; only one statement needs to be added:

```
//demo1
[code]
1    #include <stdlib.h>
2    #include <stdio.h>
3    #define LEN 5
4    int main(int argc , char* argv[])
5    {
6        int i;
7        float x=2;
8        float arr[LEN];
9    #pragma offload target (mic)
10       for(i=0;i<LEN;++i)
11       {
12           printf("Index:%d\n",i);
13       }
14       return 0;
15   }
```

```
!fortran
[code]
    1           program demo1
    2           integer,parameter ::len=5
    3           integer :: i
    4           REAL, DIMENSION(len)::arr
    5   !dec$ offload begin target(mic)
    6           do i=1,len
    7             print *,"Index:",i
    8           enddo
    9   !dec$ end offload
   10           end program
```

The "offload" command, which is the two lines with "pragma" in C code (or two lines with "!dec$" in Fortran code) tells the compiler that the subsequent segment is to be run on MIC. (The definition of "code segment" can be complicated. For simplicity, one sentence must be within the same segment. While for multiple sentences Fortran is easier to understand, all the code between one sentence and its relative end belongs to one code segment. In C/C++, it is defined by square brackets: {}. However, sometimes the grammar allows the omission of these brackets, so we need to also consider these situations. Some common cases are: loop sentence and loop body as one code segment, the sentence and body as one segment, and the sentences within the same brackets are all one segment.)

In compiling, we do not need special options.

```
icc –o demo1 demo.c
```

```
ifort –o demo1 demo.F
```

Therefore, the newest compiler will build the MIC program by default. If you only want to compile the CPU program, a "-no-offload" option is needed.

However, after execution, the result is the same as that of the CPU sequential code. Isn't MIC in parallel? If the results are the same, how can we determine whether it runs on MIC successfully?

What first needs to be clarified is that this code is run sequentially; a single hardware thread of one core is used. Therefore, the offload command itself only tells the compiler to execute this part of the code on the device end; it does not indicate the code that runs in parallel. Then how can we make the code run in parallel? Remember the OpenMP we talked about previously? Combining OpenMP on the device end, we can make the code run on the MIC. But there is no rush; before we use OpenMP on the device end, we will deal with the second problem first: How can we determine whether the code runs on the MIC successfully? MIC provides a macro definition "__MIC__". This macro is only defined when the code is running on the device end, so we just check to see if this macro is defined. However, this definition check cannot occur in the offload segment. To avoid this, we put it into a subfunction. The code to check the __MIC__ macro is the following:

```
[code]
 1  #include <stdio.h>
 2  #include <stdlib.h>
 3  #include <string.h>
 4  #define LEN 5
 5  __attribute__((target(mic))) void funcheck(int i)
 6  {
 7  #ifdef __MIC__
 8          printf("Index on MIC:%d\n",i);
 9  #else
10          printf("Index on CPU:%d\n",i);
11  #endif
12  }
13  int main(int argc,char* argv[])
14  {
15          int i;
16  #pragma offload target (mic:0)
17          for(i=0;i<LEN;++i)
18          {
19                  funcheck(i);
20          }
21          return 0;
22  }
```

```fortran
[code]
 1          program demo1
 2
 3          integer,parameter ::len=5
 4          integer :: i
 5          REAL, DIMENSION(len)::arr
 6    !dec$ offload begin target(mic)
 7          do i=1,len
 8              call funcheck(i)
 9          enddo
10    !dec$ end offload
11    contains
12
13    !dec$ attributes offload:mic::funcheck
14          subroutine funcheck(i)
15          IMPLICIT NONE
16          integer,INTENT(in)::i
17    !dec$ if defined(__MIC__)
18              print *," Index on MIC:",i
19    !dec$ else
20              print *," Index on CPU:",i
21    !dec$ endif
22          end subroutine
23
24          end program
```

The output is following:

```
Index on MIC:1
Index on MIC:2
Index on MIC:3
Index on MIC:4
Index on MIC:5
```

Readers can try the result by compiling with "-no-offload". By ignoring the declaration, we can see that by checking the "__MIC__" macro, we can determine whether the code is running or not on the device end.

5.3.1.2 Data Transfer

Although the main function of our first demo is to print output, for practical purposes, we do not suggest printing output on the device end. Of course, it is fine as a debugging tool, but it is not recommended as a common way to produce output running information, error messages, etc. It is easy to understand that when running code on the CPU side, output is done with the CPU output device (such as a monitor or hard drive), and memory might also be involved. However, when using the device output (such as the MIC card), it must relay by the CPU. Although vendors achieve this differently, the on-board cache needs to be used, transferred to the host memory, and then the output. Therefore, it has a strong impact on performance. However, hardware limitations lead to inaccuracy (mostly concerning the time sequence) or completeness (due to cache size) in the output. So we strongly recommend not doing output on the device end.

So, we are going to program this demo code for computing purposes, not for printing anything on the device end. From now on, we will only maintain MIC code, instead of two sets of codes. This is the advantage of MIC using directives to program. Whether MIC or OpenMP, those directives let a simple set of code fit different program models. Without related compiler options, we can remove features without changing the code. This is a great benefit while debugging and making the code suit different hardware. In our demo, depending on requirements, we can compile either the CPU or MIC version of the program with different compiler options.

Our code is the following:

```
//demo1
[code]
1    #include <stdlib.h>
2    #include <stdio.h>
3    #define LEN 5
4    int main(int argc , char* argv[])
5    {
6         int i;
7         float x=2;
8         float arr[LEN];
9    #pragma offload target (mic)
10        for(i=0;i<LEN;++i)
11        {
12            arr[i]=i*3.0f/x;
13        }
14        return 0;
15   }
```

```
!Fortran
[code]
   1            program demo1
   2            integer,parameter ::len=5
   3            integer :: i
   4            REAL, DIMENSION(len)::arr
   5            real::x
   6            x=2
   7    !dec$ offload begin target(mic)
   8            do i=1,len
   9              arr(i)=i*3.0/x
  10            enddo
  11    !dec$ end offload
  12
  13    end program
```

However, there is a simple difference: the program must use the array operation, so new codes must be added to make the code correct. Now we should revise the offload citation as:

```
#pragma offload target (mic) out(arr)
```

```
!dec$ offload begin target(mic) out(arr)
```

"Out" is the new keyword. This tells the compiler that the variable or array inside the brackets needs to be outputted. Then when the code leaves the MIC card, the driver copies the variable to the related position in the host memory. There are keywords similar to this: in, inout, and nocopy; later on we will explain their meanings. In addition, because the "arr" array is declared on stack, its size is fixed when it is compiled; there is no need to mark its size when it is transferred. However, if the array is declared on heap, due to its unknown size when compiled, it must input its size, when being transferred; we will explain this grammar a bit later.

An astute reader might discover a variable "x" is not passing initial value by the "in" keyword. In fact, this is one of the easy aspects of MIC programming: a non-array variable, if not passed explicitly and used by MIC, automatically passes as input. Of course, the programmer can pass this value explicitly, but in normal cases it will not have a strong impact on performance.

After examining this functionality of input and output, we can verify the correctness of our code. The revised code is the following:

```
//demo1
[code]
1    #include <stdlib.h>
2    #include <stdio.h>
3    #define LEN 5
4    int main(int argc , char* argv[])
5    {
6        int i;
7        float x=2;
8        float arr[LEN];
9    #pragma offload target (mic) out(arr)
10       for(i=0;i<LEN;++i)
11       {
12           arr[i]=i*3.0f/x;
13       }
14       if(fabsf(arr[2]-2*3.0f/x)<1e-6)
15           printf("Demo is right\n");
16       else
17           printf("Demo is wrong,arr[2] is %f\n",arr[2]);
18       return 0;
19   }
```

```
!fortran
1            program demo1
2            integer,parameter ::len=5
3            integer :: i
4            REAL, DIMENSION(len)::arr
5            real::x
6            x=2
7    !dec$ offload begin target(mic) out(arr)
8            do i=1,len
9                arr(i)=i*3.0/x
10           enddo
11   !dec$ end offload
12
13           if( abs(arr(2)-2*3.0/x) .lt. 1.0e-6)then
14                   print *,"Demo is right"
15           else
16                   print *,"Demo is wrong,arr[2] is ",arr(2)
17           endif
18           end program
```

Test this now, and if there is a "Demo is right" output, it shows our code revision is successful.

Before continuing to the next step, we will explain the meaning and usage of several keywords mentioned previously.

in: Input. Allocate memory on the device end and copy data from the host end.

out: Output. Allocate memory on the device end, and when finished running on the device end, copy the data to the host end.

inout: Input and output. Allocate memory on the device end, and when running on the device end also copy data to the device end; after running is complete, copy the data to the host end.

nocopy: No copy. Only allocate memory; do not copy the data.

At first the "nocopy" keyword might be a little bit difficult to understand. Normally, there are two places to use nocopy. One is in a different offload region (this can be interpreted as offload twice): if the next one intends to use some of the variables or data from the previous one, then nocopy can be used to avoid the waste of data transferring through the host memory. Another case is when the variable is used as a temporary variable (no need for an initial value from the host or copy back to host): then nocopy can be used to avoid data transfer.

Before seeing an example of nocopy, we need to know how to use these four keywords (known as transmission-related keywords, TRKs). The grammar is as follows:

1. There can be zero or more TRKs. When there are multiples, they can be written continuously or separated with a comma or space. The same TRK can be used multiple times in an offload sentence; however, the same variable cannot appear within the same offload sentence (even with different TRK parameters).
2. TRKs must be followed by brackets; the parameters inside the brackets are C/C++/Fortran variable names.
3. Variables should be array, pointer (just for dynamic array), or normal (scalar); multiple variables should be separated by using commas.
4. When the variable is a pointer, it must point to nonpointer variables (two-dimensional arrays are not supported).
5. When the variable is an array or pointer, it can define the start and length.
6. When the variable is a pointer, it requires the addition of ":length(len)" after the variable name ("len" is the number of elements for the array). If more than one dynamic array has the same number of elements, they can be put together. For example: in(a,b,c:length(20)). The number of elements can be a variable.
7. Besides length, there are other keywords such as alloc_if, free_if, align, alloc, and into. Before using these keywords, they must be separated with a colon (one colon for each keyword).
8. The parameters "alloc_if" and "free_if" are judging expressions with logical results. If alloc_if is true, then the memory will be allocated before entering the device end. If free_if is true, then the memory will be freed upon leaving the device end.
9. The parameter "align" is an integer. It must have an integer power of 2, meaning that the variable on the device end will align as "align".

10. The parameter "alloc" is a variable or array name, which means that it creates a specific memory space.
11. The parameter "into" is a variable or array name. However, it can only do one-to-one transfer. This means that it can copy one array from the host to another device, or vice versa. This keyword can be used with alloc, alloc_if, or free_if. However, it cannot be used with in'out or nocopy.

These definitions may seem somewhat boring because of their extreme precision, but more detailed explanations of these terms will follow later. For now, let us use some code to explain how these parameters work. Due to the only slight differences between Fortran and C/C++ in this area of grammar, no Fortran code is provided.

```
[code snippet]
//The two following lines only allocate memory on device end
#pragma offload target(mic) nocopy(p:length(sz) alloc_if(1) free_if(0))
{};
//nocopy: no need to copy from host end, and data will not copy back from device to host when the code
ends.
// p:length(sz): the array of nocopy is named as p and with sz elements \
        (Note, the length is not the array size, but the number of elements)
//p: must be declared on host end, can only declare a pointer without allocate memory.\
        If no declaration, MIC end will never know the type of p.
// alloc_if(1): allocate memory
// free_if(0): when this code segment(this offload) ends, will not free memory.


//The following two lines will copy data from host end to device end.
#pragma offload in(p:length(sz) alloc_if(0) free_if(0))
{/*Here will do computation with array p */ }
//in: copy data from host end to device end
//As p is declared in the previous offload code segment, \
        and not released so it can be used directly.
// alloc_if(0): do not allocate memory. Because it was not freed previously.


// Those two lines in the following prevent any change of allocation of p
#pragma offload nocopy(p)
{ /* Here will do computation with array p */ }
//Without any explicit speciation, not in/out/alloc/free operations,\
        only in this usage, no need for array length.


//The following two lines will output and free the memory
#pragma offload out(p:length(sz) alloc_if(0) free_if(1))
{/* Here will do computation with array p */ }
//out: copy back data from device end to host end when exit
// alloc_if(0): Do not allocate memory (not freed previously)
// free_if(1): Free when exit
```

One special case is if the transferring pointer is pointing to a static variable and the variable is declared with __declspec(target(mic)). In this situation, the alloc_if and free_if are ignored.

The previous example only explains the usage, although in practice the programmer must decide on his/her own.

For example, some of these attributes are used in the following demonstration:

```
[code]
1    #include <stdlib.h>
2    #include <stdio.h>
3    #include <omp.h>
4    #define LEN 5
5    __attribute__ ((target (mic))) float *a; //a is an intermediate variable, not needed to allocate space on
     CPU end. However, must be defined as global variable and with attribute prefix.
6    int main(int argc , char* argv[])
7    {
8              int i;
9              float b[LEN];
10   #pragma offload target(mic:0) nocopy(a:length(LEN) alloc_if(1) free_if(0))
11             for(i=0;i<LEN;++i)
12             {
13                       a[i]=i;
14             }
15   #pragma offload target(mic:0),out(b),nocopy(a)
16             for(i=0;i<LEN;++i)
17             {
18                       b[i]=i+a[i];
19             }
20             return 0;
21   }
```

The introduction of these attributes gives greater ease of programming; for example, dividing into in/nocopy/out can count time separately. A piecewise calculation can also be done to reduce the memory usage of the device end. Of course, the more useful ability of nocopy is to reduce the time for transferring files; this advantage can counteract the trouble of variable declaration. More and better usages are left for the reader to discover.

As for the in/out/inout sentence, there is a more practical syntax for transferring part of an array, shown in the following demostration code:

```
    [code snippet]
1   typedef int ARRAY[10][10];
2   int a[1000][500];
3   int *p;
4   ARRAY *q;
5   int *r[10][10];
6   int i, j;
7   struct { int y; } x;
8   #pragma offload ...  in( a )
9   #pragma offload ... out( a[i:j][:] )
10  #pragma offload ...  in( p[0:100] )
11  #pragma offload ...  in( (*q)[5][:] )
12  #pragma offload ... out( x.y )
```

After the in/out sentence, which can reference part of the array, the array dimension can be expressed by "[]". The first offload sentence (line 8) is the most common case, passing all of array a. The offload sentence in line 9 passes only part of array a; [i:j] defines the first dimension (i defines the start and j defines number of elements), and the second dimension only has [:], expressing that the second dimension is intact. So the data that will be passed is a[i][0]–a[i+j-1][499]. As shown in this sentence, the length parameters (i,j) can be variables. Line 10 directs the program to pass the array pointed to by p, 100 elements from zero. It shows that even if a pointer is used for a dynamic array, "[]" can be used to define the array dimensions. In line 11, the first dimension only has a 5, meaning that the first dimension only has one element, so this line will pass int[5][0]–int[5][9] (ARRAY is the alias of int[10][10], q is a pointer to int[10][10]). Line 12 expresses the part of the structure that can be transferred readily.

The above methods can help us transfer part of the data easily, while reducing the transferring time and minimizing code modification. While using these methods, please notice that even transferring only a part of the array will make full allocation of the array starting from the first element on the MIC card, so this method does not reduce memory usage. On the other hand, the array must be used as a whole. When ignoring the offload sentence, or assuming that the code is running on the CPU end, the code should stay the same when passing part of the array. This is designed to avoid maintaining two sets of code. For example: Assume there is an array p[100], while in(p[2:10]), when used on the MIC end, it will allocate a 12-element space(p[0]–p[11]) and will copy p[2]~p[11] from the host memory to the MIC end memory, thus the first effective element is still p[2] instead of p[0].

There are two keywords according to this usage, alloc and into.

As said above, the grammar that passes part of the array creates the whole memory space. In some cases the whole space may not be needed, so alloc can be used to create a limited space. For example:

```
[code snippet]
#pragma offload ... in ( p[10:100] : alloc(p[5:1000]) )
```

The offload line first creates a 1000-element array p; the available range is from 5, namely 5–1004. It then passes 100 elements from the host starting from p[10], namely p[10]–p[109], to the device end position p[10]–p[109]. However, the overflow check is still the programmer's responsibility.

The into sentence can pass part of a host array to a device array, and vice versa. For example:

```
[code snippet]
#pragma offload ... in( p[0:500] : into (p1[500:500]) )
```

This offload line copies 500 elements on the host, starting from p[0], to the related position p1[500]–p1[999] on the device end.

When using this method, the programmer needs to control the correctness, especially when there is overlap, such as:

```
[code snippet]
#pragma offload ...      in( p[0:600]    : into (p1[0:600]) )       \
                         in( p[601:400] : into (p1[100:400]) )
```

Here, the object array p1 has an overlap between two transfers; one is 0–599, the other is 100–499. The overlap happens in 100–499; this will cause a "no definition" error, namely that in one offload sentence, the running order is not certain.

"into" is not a simple memory copy, so it can't pass data between arrays with different dimensions. For example:

```
//ERROR!
[code snippet]
int rank1[1000], rank2[10][100];
#pragma offload ...      out( rank1 : into(rank2) )
```

rank1 and rank2 have different dimensions, so they can't pass directly.

5.3.1.3 Target

Let us now return to the offload sentence. There are a few more keywords left, so let us finish them.

First, target(mic): "mic" is the only available value at the moment. The keyword "target" can point out which MIC card it will use. "mic" is always followed by a colon and a number, such as target(mic:1). If the number is −1, the system will choose a device automatically (right now, only among MIC cards); if there is no available device (i.e., we only have the CPU), the program will exit and report an error. If the number is equal to or greater than zero, the program will offload it to the specified device; the device number = (number) mod (number of devices). Numbers less than −1 are reserved for system usage and are unavailable.

If no MIC card is specified for a situation where there is no MIC card, the program will run on CPU and no error will be reported. When there is more than one MIC card, it will use multiple cards in turn. Therefore, if a MIC card is not specified, the programmer must pay special attention when running a performance test. These cases will interfere with the test result:

1. When the MIC card malfunctions, the program runs on the CPU but mistakenly believes that it is running on MIC.
2. When there are two or more MIC cards (due to the first run of offload, the MIC needs initialization; usually the time does not count in total running time), the second offload can be mistakenly counted as the second run on the first card when it actually is the first run on the second card.
3. When there are two or more MIC cards of different models; this will lead to some differences in the performance. The performance of the program could falsely appear as being unstable.

There is a great deal of freedom in using multiple MIC cards to perform parallel computing. It can use pThread to define different MIC card numbers in thread functions and to add another layer of an OpenMP loop. The OpenMP version of the code is as follows:

```
[code snippet]
//C/C++
1    omp_set_nested(1);
2    #pragma parallel for num_threads(3)// Assume there are 2 MIC cards, collaborating with CPU
3    for(i=0;i<3;++i)
4    {
5    #pragma offload target(mic:i) if(i>1) in(...) out(...)
6    ...
7    }
```

```
[code snippet]
!Fortran
1    call omp_set_nested(.true.)
2    !$omp parallel do num_threads(3) ! Assume there are 2 MIC cards, collaborating with CPU
3    do i=1,3
4    !dir$ offload target(mic:i)   if(i>1)   in(...) out(...)   !if grammar see following
5    ....
6    enddo
```

The above code distributes the first task to the CPU, and the second and the third tasks to two different MIC cards. Due to the usage of the CPU and the fact that the

CPU code is also OpenMP, a nested OpenMP code block is formed. Here, we need to call omp_set_nested(.true.) to make the program able to run nested OpenMP; otherwise, only the outer loop will run in parallel.

Before offload, we can use the API function int _Offload_number_of_devices (void) to get the number of MIC devices in the system. In the offload code block, we can use the API function int _Offload_get_device_number(void) to obtain the number of devices this code block is running on.

5.3.1.4 The If Statement

There is an "if" statement in OpenMP, which can decide if it is going to be paralleled or not according to different conditions. There is also"if" usage in offload to decide if the code block is going to be put on the device. If the result of the "if" expression is false, the code block will run on the CPU; otherwise, it will run on the MIC end. For example: #pragma offload target(mic) if(N>1000), means that if N is greater than 1,000, use MIC, otherwise use CPU.

Through "if", the programmer can freely select the environment that uses MIC. In some low-parallelism applications, we can choose not to use MIC, while we would choose to utilize it in a highly parallel environment. However, this free adjustment does not require editing or recompiling the code.

There are many ways to use "if": besides choosing the appropriate device depending on the computational scale, it can also be used in multi-card computing as mentioned above, deciding which card the flag bit should be run on.

5.3.1.5 Mandatory

There is another keyword for offload: "mandatory". If the mandatory keyword is in the offload line, then this code block must be run on MIC. If MIC is unavailable, it cannot be run on the CPU, so it will report an error and instead will exit. We note that mandatory cannot be use with "if" statements simultaneously.

5.3.1.6 Asynchronous Transmission

"Signal" and "wait" belong to asynchronous transmissions; the CPU end does not need to wait for the offload command to return before running the code. It normally uses parallel execution of the CPU code after launching the MIC code block in order to execute them at the same time. Another way is to use the offload_transfer and offload_wait; these two are similar to offload, but are only in charge of data transfer, not computing code. The offload_transfer supports the same parameters as offload, but offload_wait only supports three parameters: target, if, and wait. The usage of signal and wait is the same in two different ways. Signal sends a signal after the offload block is finished, and wait receives it; these two must be used as a pair. However, one wait can wait for multiple signals at the same time, so the numbers of signal and wait may not match. The parameter "tag" for signal and wait, which passes one of the array pointers in the C language (namely, one of the array names in in/out/inout), can only signal/wait for one array when transferring multiple arrays. In Fortran, it is an integer variable for identification purposes. For example, we have:

```
[code snippet]
//C/C++
1    int counter;
2    float *in1;
3    counter = 10000;
4    __attributes__((target(mic))) mic_compute;
5    while(counter>0)
6    {
7    #pragma offload target(mic:0) signal(in1)
8    {
9        mic_compute();
10   }
11   cpu_compute() // This function will run parallel with MIC function above
12   #pragma offload_wait target(mic:0) wait(in)
13   counter--;
14   }
```

```
[code snippet]
!Fortran
1    integer signal_var
2    integer counter
3    counter = 10000
4    !DIR$ ATTRIBUTES OFFLOAD:MIC :: mic_compute
5    do while (counter .gt. 0)
6        !DIR$ OFFLOAD TARGET(MIC:0) SIGNAL(signal_var)
7          call mic_compute()
8        call cpu_compute() !This function will run parallel with MIC function above
9        !DIR$ OFFLOAD_WAIT TARGET(MIC:0) WAIT (signal_var)
10       counter = counter - 1
11   end do
12   end
```

In this example, we define a MIC computing function mic_compute and a CPU computing function cpu_compute. When the program runs into offload, the MIC end will run mic_compute and hand over control to the CPU thread. After the CPU gains control, it will execute the cpu_compute function. After cpu_compute is finished, it will execute offload_wait, by checking whether the signal of the previous offload has arrived. If not, it will wait; otherwise, it will continue to execute the following command, counter self-minus.

For more examples, the reader should go to Chap. 8.

5.3.1.7 Summary of Offload Grammar

Now, let's summarize the usages of offload with examples in C/C++:

```
#pragma offload specifier[, specifier...]
The options which specifier can fill in are:
   target          eg: target(mic:0)
   if        eg: if(N>1000)
   in              eg: in(p:length(LEN) alloc_if(1))
   out             eg: out(p:length(LEN))
   inout           eg: inout(p:length(LEN) align(8))
   nocopy          eg: nocopy(p)
   signal          eg: signal(tag)
   wait            eg: wait(tag1,tag2)
   mandatory       eg: mandatory
And the available properties of in/out/inout/nocopy are:
   length          eg: :length(LEN)
   alloc_if        eg:   :alloc_if(1)
   free_if         eg:   :free_if(N>0)
   align           eg:   :align(8)
   alloc           eg:   :alloc(p[10:100]) , cannot work together with inout/nocopy
   into            eg:   :into(p[10:100]), cannot work together with inout/nocopy
```

For Fortran, just change the #pragma to !DIR$.

5.3.1.8 Offload Combined with OpenMP

Although we have learned how to place code segments that run on MIC, the program is still single-threaded, and the overall advantages of the multi-cores in the MIC card have not properly been displayed. At this point, we need to use OpenMP. The usage is very simple: add in the statement block offload indicated, and then we can use OpenMP. In Fortran, for example:

```
[code snippet]
!DIR$ OMP OFFLOAD TARGET(mic)
!$omp parallel
.......
!$omp end parallel
```

Now, our example has been updated to the form below:

```
[code]
//demo1
1      #include <stdlib.h>
2      #include <stdio.h>
3      #define LEN 5000
4      int main(int argc , char* argv[])
5      {
6          int i;
7          float x=2;
8          float *arr=(float*)malloc(LEN*sizeof(float));
9      #pragma offload target (mic) out(arr:length(LEN))
10     #pragma omp parallel for
11         for(i=0;i<LEN;++i)
12         {
13             arr[i]=i*3.0f/x;
14         }
15         if(fabsf(arr[2]-2*3.0f/x)<1e-6)
16             printf("Demo is right\n");
17         else
18             printf("Demo is wrong,arr[2] is %f\n",arr[2]);
19         return 0;
20     }
```

```
[code]
!fortran
1           program demo1
2           integer,parameter ::len=5
3           integer :: i
4           REAL, DIMENSION(len)::arr
5           real::x
6           x=2
7      !dec$ offload target(mic) &
8      !dec$ out(arr)
9      !$omp parallel do
10         do i=1,len
11             arr(i)=i*3.0/x
12         enddo
13     !$omp end parallel do
14             if( abs(arr(2)-2*3.0/x) .lt. 1.0e-6)then
15                     print *,"Demo is right"
16             else
17                     print *,"Demo is wrong,arr[2] is ",arr(2)
18             endif
19         end program
```

In this way, we can use OpenMP and take full advantage of the MIC multi-cores. Please do not forget to add the –openmp to the compiler options. The exact number of MIC cores and threads that need to be used also depends on the OpenMP related functions.

5.3.2   Declarations of Variables and Functions

Having introduced only the offload statement, we are now ready to use MIC for high-performance computing. However, only using offload will cause a software engineering problem. Functions in the original codes cannot be used. So we must employ a grammar for declaring functions to ensure that the MIC program can use these functions.

For this kind of declaration only, we need to add one declaration to the existing code:

```
C/C++:
__declspec( target (mic)) declarations of variables and functions
or
__attribute__ (( target (mic))) declarations of variables and functions
```

```
Fortran:
!DIR$ attributes offload: target-name :: routine Name or variable name [ , routine Name or variable name] ...
```

For example:

```
[code snippet]
C/C++:
__attribute__ (( target (mic))) int a;
__attribute__ (( target (mic))) void func();
```

```
[code snippet]
Fortran:
!DIR$ attributes offload:mic:: a,func
real :: a
subroutine func(var)
...
end subroutine
```

The grammar here is very simple. Only two points need to be emphasized: first, in C/C++, if we use this attribute, we only need to remember that there are two layers of brackets in the target periphery; only writing one layer will cause a grammar error. Second, the declaration can be used for a function or a variable; when used on a variable, the variable must be global. Once declared, functions and variables can be used for both CPU and MIC codes.

If there are a lot of variables or functions, it is inconvenient to add definitions for each and every one. Luckily, MIC provides a batch declaration, so we can declare multiple functions or variables at one time, as well as mix declaration functions and variables.

```
C/C++:
#pragma offload_attribute([push, ] target(target-name))
    //Declarations of variables and functions
#pragma offload_attribute(pop|{target(none)})
```

```
Fortran:
!DIR$ OPTIONS /OFFLOAD_ATTRIBUTE_TARGET=mic
    !Declarations of variables and functions
!DIR$ END OPTIONS
```

There are two ways for using C/C++, as shown below:

```
[code snippet]
#pragma offload_attribute (push, target (mic))
// Declarations of variables and functions
#pragma offload_attribute (pop)
```

or:

```
[code snippet]
#pragma offload_attribute (target (mic))
// Declarations of variables and functions
#pragma offload_attribute (target(none))
```

For example:

```
[code snippet]
#pragma offload_attribute (push, target (mic))
    int a;
    float func();
#pragma offload_attribute (pop)
```

```
[code snippet]
!DIR$ OPTIONS /OFFLOAD_ATTRIBUTE_TARGET=mic
    real:: a
    subroutine func(var)

    ...

    end subroutine
!DIR$ END OPTIONS
```

5.3.3   Header File

In C/C++, if using the API (to get the number of devices, for example) of MIC, we need to contain offload.h.

In Fortran, if using the API of MIC, we need to include mic_lih.f90 or USE mic_lib.

5.3.4   Environment Variables

The device program of MIC provides some environment variables. Using them as needed makes program execution or debugging more convenient. We introduce some commonly used environment variables below; for the completed or updated version, please refer to the compiler manual.

5.3.4.1 MIC_STACKSIZE

MIC_STACKSIZE defines the stack size of each thread on MIC. Its default value is 2 MB. Because 2 MB may not be enough, it can be changed accordingly by this environment variable. The units it supports are B, K, M, G, and T. For example: export MIC_STACKSIZE=5M.

5.3.4.2 MIC_ENV_PREFIX

MIC_ENV_PREFIX sets the prefix of the environment variables belonging to MIC. This distinguishes between the environment variables on the MIC end and those on the CPU end. If this distinction is not made, the CPU-end environment variable will also be used on the MIC end by default. For example, if we define OMP_NUM_THREADS=8 on the CPU end, OpenMP can launch 8 threads on the CPU. However, then on the MIC end, OpenMP can only launch 8 threads also, which is not what we would expect. Therefore, we must distinguish the environment variables between those two ends. For example:

```
export   OMP_NUM_THREADS=8
export   MIC_OMP_NUM_THREADS=124
export   MIC_ENV_PREFIX=MIC_
```

Now, on the CPU end, the number of OpenMP threads is 8, while on the MIC end, the number of OpenMP threads is 124.

The way MIC_ENV_PREFIX works is that it defines a prefix (such as "MIC_"). If there are environment variables that fit the description (such as "MIC_OMP_NUM_THREADS), then remove the prefix and apply them on the MIC end. Namely, the environment variable is still "OMP_NUM_THREADS", but its value is defined by "MIC_OMP_NUM_THREADS".

5.3.4.3 MIC_LD_LIBRARY_PATH

MIC_LD_LIBRARY_PATH defines the shared library path for the MIC program. Here, the path refers to the MIC card address. It requires copying the shared library to the MIC end. MIC_LD_LIBRARY_PATH normally points to the user-defined shared libraries because system libraries have the default path (/lib64), so normal users can't get access to copy files to that path. This is different from other environment variables because MIC_LD_LIBRARY_PATH does not need to be used with MIC_ENV_PREFIX since the host end LD_LIBRARY_PATH will not be extended to the MIC end.

5.3.5   Compiling Options

There are not many differences between MIC and CPU programs, and most of the CPU options are also available on the MIC. Some of the most useful options are mentioned above, but here we give a summary of MIC-related compiler options.

5.3.5.1 Mmic

-mmic expresses that the program can only run on MIC cards. While using this option, the compiler will define the macro __MIC__. This option is disabled by default and will compile a heterogeneous program while it is disabled.

5.3.5.2 No-offload

-no-offload expresses that the program can only run on the CPU end. This option will ignore all MIC-related lines, such as offload. It is disabled by default.

5.3.5.3 Offload-Attribute-Target

The way to use this option is -offload-attribute-target=mic, which expresses that all MIC-usable functions and variables can be used on MIC. It has the same effect when using "attribute" to declare a function or variable. This is disabled by default, and all functions or variables cannot be used on MIC (namely offload command) if not declared explicitly.

5.3.5.4 Offload-Option

Used as -offload-option, target, tool, "option-list", this command is aimed at offloaded objects (i.e., the code block running on MIC) and uses special compiling options. For example, "target" can only choose mic, and "tool" can use either ld, as, or compiler. The option-list is for specific options, which must be put inside quotes, and the options must be separated by a space.

When compiling heterogeneous programs, all compiling options will be passed through the host end compiler; thus, part of the options (MIC supported) will be passed to the MIC-end compiler. To see which options have been passed, use "-watch-mic-cmd".

The -offload-option adds compiling options that can only be used on the MIC end; it will overwrite or attach to the CPU-end options. If you only want to override the autotransfer options, -offload-option must be put before the autotransfer options; if -offload-option is put afterward, it will automatically add autotransfer options after. These are a few examples:

```
-offload-option,mic,compiler, "-O3 -diag-disable 1234 –vec-report2"
```

This demonstrates how to pass options to the compiler:

```
-offload-option,mic,ld,"-lmylib -L/my/path"
```

The following demonstrates how to pass options to ld, where ld is the linker for libraries. Note that the path it passes is normally the shared library path (on-card), similar to the required library names when compiling with -llib while defined on the complier part.

```
-offload-option,mic,as,"<assembler options>"
```

This demo shows how to pass options to as, where as is the assembler.

```
icc  -offload-option,mic,compiler, "-I/my_dir/include -L/my_dir/lib -DMY_DEFINE=10 -I/my_dir/mic/include
-DMY_DEFINE=20" -offload-option,mic,ld, "-L/my_dir/mic/lib" hello.c
```

This demo shows a full compile command. The program will use -DMY_DEFINE=20 instead of 10, because the options that follow after will override the previous options. For the include path, it will search /my_dir/mic/ include prior.

```
-offload-option,mic,compiler,"-O3 -DMY_DEFINE=MIC" -DMY_DEFINE=HOST -O2
```

In contrast to the previous demo, -O3, -DMY_DEFINE=MIC has more priority here. It will be used on the MIC end instead of O2 and HOST being passed by the CPU end. However, for the CPU-end program, it uses O2 and HOST. Thus, the CPU and MIC ends use different options independent of each other.

5.3.6 Other Questions

The offload region can only be exited in the normal way. If we use exit() in offload, it will cause an "EventWait failed: COI_PROCESS_DIED" error.

In Fortran, the variables defined in the external function or subroutine cannot be passed to the internal function or subroutine through the offload statement except for SAVE variables, which cannot access the variable host subroutine variable in the internal subroutine offload block. If this is needed, please use the temporary variable "transit".

At this point, we have introduced the common MIC grammar. With the above commands, which are sufficient for the multi-core characteristics of the MIC, we can now write high-performance computing programs.

5.4 MPI on MIC

There is no difference in using MPI on MIC or the CPU. This section only gives a simple introduction. For more details, please read related books and manuals. In addition, this section covers the unique method to set the MPI running environment on MIC, which is a little different from the CPU.

5.4.1 MPI on MIC

This section focuses on the Intel intelligent platform management interface (IMPI) 4.1.0.024. The Intel MPI supports the Intel® Xeon Phi™ coprocessor based on MIC. All the MIC cards we mentioned in this section denote the Intel® Xeon Phi™, named Knights Corner. IMPI supports communication of CPUs as well as of MIC, including some special MIC communication functions.

Supported features include:

- communication on MIC
- communication between MIC and the host CPU for a single node
- communication between MICs for a single node
- communication between MIC/CPU for multi-nodes

Supported compilers include:

- Intel C++ 13.0 or higher version (for Linux)
- Intel Fortran 13.0 or higher version (for Linux)

The features temporarily not supported include:

- Task manger MPD
- Checkpoint-Restart

Please see the official MPI document at http://software.intel.com/en-us/intel-mpi-library/ for more detailed features and future versions.

5.4.2 MPI Programming on MIC

The MPI programming mode on MIC is divided into two groups: offload mode and native mode.

Under the offload mode, there are two situations to make the MIC or CPU an accelerator: CPU as the host or MIC as the host. For the situation of the CPU as the host, MPI ranks only exist on the CPU host, which passes messages while MIC is only in charge of acceleration. In contrast, when MIC is the host, MPI rank only exists on the MIC host and messages are passed by MIC while CPU is an accelerator.

Under the native mode, MIC and CPU are peer nodes. Here, there are also two situations: MPI ranks on MIC only, or MPI ranks on both MIC and the host CPU. In the first situation, only MIC can fetch MPI ranks, the messages go into and out of MIC, and code threading is possible. In the second situation, MPI ranks exist on MIC and the host CPU, so both MIC and the CPU can fetch MPI ranks and code threading is also possible. These modes are shown in Fig. 5.2.

There are five combinations on how computational and MPI functions can run on CPU or MIC devices, as shown in Fig. 5.3.

The CPU-native (Xeon®-native) mode puts processes and MPI functions only on the CPU; meanwhile, MIC does no work. This was the most common mode before the birth of MIC.

The CPU-hosted offload mode (Xeon® hosted MIC coprocessed mode) is similar to the current programming mode of GPGPU. In this case, MPI processes only exist on the CPU. The CPU controls MIC using the offload mode, so that the MIC is treated as the coprocessor and there are only MPI communications between CPUs. This mode requires the least rewriting of existing codes. It is also the only solution for communication between multi-nodes. The CPU-hosted mode is described in detail in Fig. 5.4.

From Fig. 5.4, we can see that the CPU controls the MIC coprocessor (dark hollow arrows) with the offload method. MPI messages (light hollow arrows) are transported, and data can access the network through CPUs (solid arrows cross the CPU).

Under the MIC-hosted offload mode, MPI processes only appear on MIC. The CPU, as the coprocessor, is called by MIC with the offload method and does not join the MPI in managing and communicating. There are only MPI communications between the MIC cores or MIC cards, with no communications across the different nodes.

Native mode takes MIC as an independent node. Each MIC core can be considered as a node. In this case, the MIC–MPI program can be divided into two groups: MPI program running on MIC only and running on MIC/CPU together. In both groups, MIC can be taken as a node, with messages passing on nodes (regardless of whether the MPI processes are on the CPU or MIC). This MPI calling

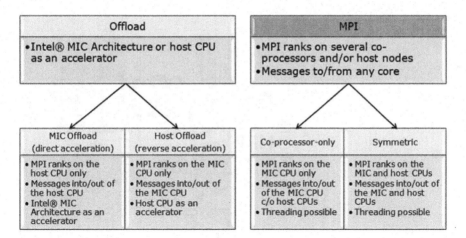

Fig. 5.2 MPI programming modes on MIC

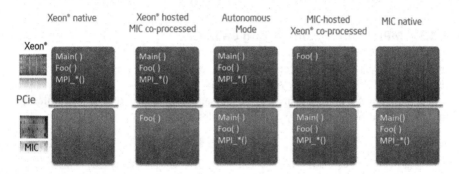

Fig. 5.3 MPI execution modes on MIC

method seems no different from the traditional CPU-based calling method. Figure 5.5 shows the native mode of the MPI program running on MIC, and Fig. 5.6 shows the program running on the MIC/CPU together.

MPI messages (bold arrows) can only be passed in the inner node for both modes. The only difference is in which device the MPI is messaging. Three possible messaging methods are: through the MIC cards, through the CPU, and through both the CPU and MIC. If communicating with other nodes in the network is necessary, messages must be passed through the CPU. See the thin arrows in the figures, which go across the CPUs.

It can be simply understood that a virtual local network combined by MIC and CPU has a unique IP. But this IP is not addressed and routed together with nodes by the CPU. Therefore, the communication cannot cross different nodes by using MIC.

Fig. 5.4 Running MPI with offload mode

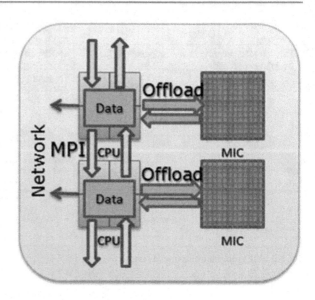

5.4.3 MPI Environment Setting on MIC

First, we must have the root access and make sure that OpenSSH is installed in the server. The SSH item "Pubkey Authentication" must be set to "yes".

Then, make sure that there are id_rsa, id_rsa.pub, id_dsa, and id_dsa.pub files in the .ssh of your home directory. Otherwise, please execute:

```
$ ssh-keygen –t rsa
$ ssh-keygen –t dsa
```

Afterwards, hit the "enter" key to keep the default setting when selecting.

Every user needs to execute the two steps mentioned above. Otherwise, the user cannot log into the MIC card. Then, execute the following commands to set a no-password login:

```
(host)$ sudo service mpss stop
(host)$ sudo micctrl --resetconfig
(host)$ sudo service mpss start
```

Some information will be written into related profiles in running micctrl. Please check the log file in case of errors.

Next, we can test the MIC. For example:

```
(host$) sudo  ssh   –x   172.31.1.1   uname –a
```

IP 172.31.1.1 is the default address for the first MIC card. MIC has some default naming rules: the IP address starts with 172.31, the third segment gives the number

Fig. 5.5 MPI program
running only on MIC

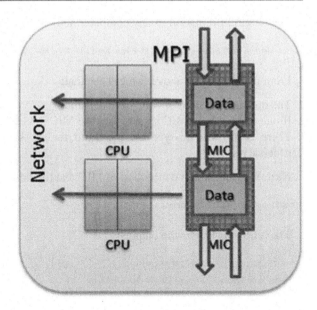

Fig. 5.6 MPI programs
running together on
MIC/CPU

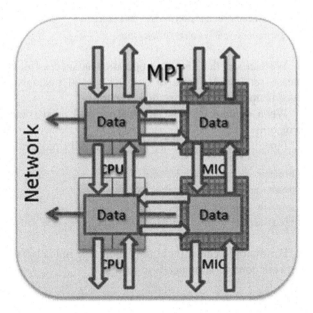

of the MIC card, the fourth segment 1 is the MIC address, and 254 is the server
address. To access MIC from the host, one can use 172.31.1.1; to access the host
from MIC, one can use 172.31.1.254. One can execute the ifconfig command for
checking the host IP address to deduce the MIC IP. All the IP addresses are
changeable. See also the MIC driver installation document for details. One can
change the network segment in the file of /etc/sysconfig/mic/default.conf.

The output information of any computer node is given as the following:

```
Linux micnode-mic0 2.6.34-gdc02fc4 #2 SMP Wed Jun 13 17:06:07 PDT 2012 k1om GNU/Linux
```

From the operations above, we can see that:

1. The operating system of MIC is GNU/Linux.
2. If one can log in to MIC by ssh, then the MIC device works.
3. There is no need to type the password for ssh login for MIC through the necessary settings.

Next, we can test the connection of MIC and the host (for the second card):

```
(host$) sudo ssh -x 172.31.2.1 ping -c 1 -w 10 172.31.2.254
```

The following shows the outputs:

```
PING 172.31.2.254 (172.31.2.254): 56 data bytes
64 bytes from 172.31.2.254: seq=0 ttl=64 time=0.849 ms

--- 172.31.2.254 ping statistics ---
1 packets transmitted, 1 packets received, 0% packet loss
round-trip min/avg/max = 0.849/0.849/0.849 ms
```

We can see that the ping operation is successful. If there are still problems, please set iptables first, or see the related manuals (driver's readme file and documents in the MPI installed directory).

When configuration is finished, we still need some preparations to run MPI programs.

First, upload some library files for the MPI program to MIC.

```
(host)$ cd    <installdir>/mic/bin/
(host)$ sudo scp   *   172.31.1.1:/bin
(host)$ cd    <installdir>/mic/lib/
(host)$ sudo scp   *   172.31.1.1:/lib64
```

For multiple cards, these files should be uploaded to everyone, and a host-end network configuration needs to be set:

```
(host)$ sudo sysctl -w net.ipv4.ip_forward=1
```

Meanwhile, one should ensure MPSS is set to p2p, which is the default mode for startup. There is no need to modify it. To configure, see /etc/modprobe.d/mic.conf. If p2p=0, please modify it to p2p=1 or just delete it. Execute the following commands to reboot:

```
user_prompt> sudo service mpss stop
user_prompt> sudo service mpss unload
user_prompt> sudo service mpss start
```

5.4.4 Compile and Run

This procedure is similar to compiling traditional MPI-CPU codes on the CPU by using mpiicc/mpiifort commands. For codes on MIC (manually uploaded MIC programs), the "-mmic" option should be added. After compiling is finished, one needs to upload the executable programs, necessary libraries, and data to the MIC card by using scp.

In the runtime, the IP addresses of the CPU and MIC should be assigned using commands such as the following:

```
mpiexec.hydra –host 172.31.1.254 -n 4 -env OMP_NUM_THREADS 4 ./test.exe.host : \
-host 172.31.1.1 -n 2 -env OMP_NUM_THREADS 16 -wdir /tmp /tmp/test.exe.mic
```

Here, mpiexec.hydra is the program to run the MPI. On the CPU, mpiexec is used for linker. The option -host denotes the node IP to start the MPI program, -env denotes the environment variable, and ./test.exe.host denotes the path of MPI program to run. Commas are used to separate different nodes. Because the path on MIC is different from that on the CPU, the option –wdir changes the running path. To run the MIC program, one needs to switch the path.

5.4.5 MPI Examples on MIC

Next, we give an example on how to use MPI on MIC, with an assumption that the MPI environment on MIC has already been set up. A simple code is the following:

```
    [code]
1   #include <mpi.h>
2   #include <stdio.h>
3   int main(int argc,char* argv[])
4   {
5           int npes,myrank,data;
6           MPI_Status mstatus;
7           MPI_Init(&argc,&argv);
8           MPI_Comm_size(MPI_COMM_WORLD,&npes);
9           MPI_Comm_rank(MPI_COMM_WORLD,&myrank);
10          if(myrank==0)
11          {
12                  data=100;
13                  MPI_Send(&data, 1, MPI_INT, 1, 0, MPI_COMM_WORLD);
14          }
15          else
16          {
17                  data=200;
18                  MPI_Recv(&data, 1, MPI_INT, MPI_ANY_SOURCE, MPI_ANY_TAG,
    MPI_COMM_WORLD, &mstatus);
19          }
20          printf("From process %d out of %d,Send/Recv: %d\n",myrank,npes,data);
21          gethostname(name,&len);
22          printf("hostname:%s\n",name);
23          MPI_Finalize();
24          return 0;
25  }
```

The above code is not complex. It starts with the initialization of MPI and then gets the size and rank of each MPI process. The variable data within rank 0 is set as 100 and is sent to the process of rank 1. For rank 1, the process sets the variable "data" as 200 and overwrites it with received information from other ranks. Then each process outputs the result. The first parameter is the rank ID, the second one is the total number of processes, and the third one is the value of the variable "data". For the third parameter, the variable "data" should be 100 all the time; otherwise, the received information is not correct. Finally, we should close and release the MPI. To verify that the MPI codes can run on different nodes, the "gethostname" function is utilized to output the hostnames.

The compiling command is the following:

```
mpiicc  mpi_helloworld.c  -o  mpihw
```

The command on the CPU is:

```
mpiexec.hydra -n 2 ./mpihw
```

And the outputs are:

```
From process 0 out of 2,Send/Recv: 100
hostname:inspur4
From process 1 out of 2,Send/Recv: 100
hostname:inspur4
```

This program can only start two processes because MPI_Send must match MPI_Recv. If multiple processes have started, only the process "rank=0" sends a message to the process "rank=1" (see the fourth parameter of MPI_Send). The process "rank>1" will hang on, because it only has MPI_Recv and won't send messages to other processes. For example, on the node "inspur4":

```
# mpiicc  mpi_helloworld.c  -o  mpihw_mic  −mmic
# scp  mpihw_mic  172.31.1.1:/tmp
```

Try to execute the following commands on MIC:

```
# ssh 172.31.1.1
# cd /tmp
# mpiexec.hydra -n 2 ./mpihw_mic
```

The output is:

```
From process 0 out of 2,Send/Recv: 100
hostname:inspur4-mic0
From process 1 out of 2,Send/Recv: 100
hostname:inspur4-mic0
```

We can see that the program runs correctly on MIC, and the node of MIC is named as "inspur4-mic0", which is the same as the hostname.

Next, we try to start MPI processes on the MIC and the CPU at the same time to communicate across nodes. Here, we take the MIC and CPU as separate nodes.

```
#mpiexec.hydra  −host  172.31.1.254  −n  1  ./mpihw :  \
-host 172.31.1.1  −n  1  -wdir  /tmp/mpihw_mic
```

The result is the following:

```
From process 0 out of 2,Send/Recv: 100
hostname:inspur4
From process 1 out of 2,Send/Recv: 100
hostname:inspur4-mic0
```

We can see that the MPI process runs on different nodes respectively, and communication crosses different nodes. Another way of starting an MPI process on two MIC cards at the same time is also possible, but make sure to upload the library and program to the second MIC card.

5.5   SCIF Programming

5.5.1   What Is SCIF?

SCIF is an acronym for Symmetric Communications Interface, which represents a simple, high-performance interface for symmetric communications. SCIF supports data transfer from the coprocessor "Knights Corner" to the CPU through the PCIE bus. SCIF gives the best performance, but some special instructions should be added to the codes because SCIF is a low-level interface close to the hardware. Thus, SCIF programming belongs to system programming.

SCIF includes MPI communications between the CPU as a host and the coprocessors. Messages can be passed among the coprocessors, the inner host node, and between the coprocessors and host node.

Figure 5.7 shows the data transfer using SCIF and offload mode, respectively. For the offload mode, data should be transferred back to the CPU for communication between different nodes. For the SCIF mode, different MIC cards can communicate directly, so transferring data using SCIF mode rather than offload mode is much more efficient. Now, we will introduce the concepts and communication principles of SCIF mode.

5.5.2   Basic Concepts of SCIF

This section covers the basic concepts below:
1. Node

The SCIF node is a physical node in the SCIF network. Both the host device and MIC can be considered to be SCIF nodes. If all the host processors share a common operating system, we can consider them as host-end SCIF nodes. In the SCIF sections of this book, we simplify "SCIF node" to "node".
2. Port

The SCIF port is a logical port of the SCIF node. A SCIF port can be described using a 16-bit integer to simulate an IP port. For example, SSH works on port 22 through a TCP connection. Each accessing port of the SCIF network is unique, and each port has a unique ID. We also simplify "SCIF port" as "port" in the SCIF sections of this book.
3. Endpoint

The connected port is called "endpoint". In general, the communication principles of endpoint are similar to the socket, although they are different conceptually.

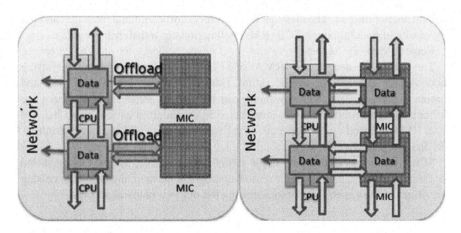

Fig. 5.7 Comparison of data transfer using offload mode (*left*) and SCIF mode (*right*)

4. Registered memory

   Registered memory is a registered address section managed by the SCIF driver.

   Register memory space can be a section of physical memory. Each connected endpoint has a registered memory.

5. Registered window

   There are limited registered addresses in the memory space, so some special sections of registered addresses are named "Registered window", or just "Window". Without such a registered window, accessing the registered memory will be rejected.

6. Pinned memory

   Pinned memory is a fixed or nonpaged memory, in which the operating system ensures the memory in the physical memory at all time. The operating system can make applications access the physical memory safely because this section of physical memory will not be broken or readdressed.

   The principle behind the pinned memory is to ensure that allocation and deallocation is working in the physical memory without paging, so the system performance is improved. Pinned memory should be applied appropriately because the physical memory is limited. In addition, SCIF also uses pinned memory to operate RMA, though the concept "windows" is utilized instead of pinned memory.

7. DMA

   Direct memory access (DMA) supports a fast transfer mechanism. At the hardware level, it allows I/O data transfer directly between devices and host memory without the processor's management. In this way, the throughput can be improved significantly. The SCIF's RMA data transfer is based on DMA.

8. Message layer

   When two endpoints are connected, messages are always transported from the local to the remote endpoint through the Message Layer. This is a two-way transport. Messages can be a random sequence of characters. Message length or content are not limited. The message layer prefers short command messages and not big blocks of data, although SCIF RMA is functionally transferring big data.

9. Remote Memory Access

The remote memory access (RMA) of SCIF supports one-way data transferring mode, which is good for I/O operation. Data can be transferred with local and remote addresses. This one-way mode can be used in some algorithms that are difficult to synchronize (such as distributed matrix multiplication) or to balance load. RMA is based on the concepts of registered memory, DMA, and so forth.

10. Remote mapping

SCIF allows remote mapping in the virtual address space, which is mapped into the threads' physical address of the remote node. Once the mapping is built, accessing the virtual address is equivalent to accessing the mapped physical address.

5.5.3 Communication Principles of SCIF

In this section we introduce the communication principles behind SCIF. "Communication" in this context means the process of sending messages. For the SCIF mode, information is transported between two endpoints. In this messaging process, one endpoint can listen to the connected endpoint while the other endpoint waits for the connection. There must be a connection before two endpoints communicate. The connection of different endpoints is similar to socket programming. Therefore, you can understand this principle easily if you are familiar with socket programming. Figure 5.8 shows the messaging process of connecting two endpoints.

Next, we introduce how the two endpoints communicate.

1. Generate new endpoint

SCIF uses scif_open() to generate new endpoints. Both ends of the communication use this function to generate a new endpoint (see Sect. 6.5.4 for details).

2. Bind address

For listening to the connection's requests, each endpoint needs to know the new generated endpoint's port ID and bind the address to this ID. The function scif_bind () is used to bind the generated endpoint to its ID. Once the address is bound to the port ID, the endpoint cannot be used for any other communication. The port address is optional; a free port will be chosen if one is not specifically named.

3. Connect

After binding the address, the endpoint can be used for listening to the connection requests both in and out of the node. However, it also needs an important process: building a virtual connection between two endpoints. The connection into the node can only be based on non-connected endpoints, so no connection can be built on the connected endpoints or on the endpoints listening into the node connections.

4. Listen

When an endpoint is bound to an address, it can be used to listen to the connection requests of the binding address.

Fig. 5.8 Flowchart of endpoint messaging

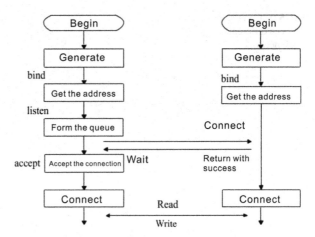

The accepting operation works on the listening endpoint. First, the accepting operation is transported into the supported protocol layer in the case of accepting any requested connection. On the other hand, the accepting operations are transported into the real protocol. The accepting operations can be blocking or non-blocking. In the situation of non-blocking, the accepting fails without an acceptable connection, and the generated endpoint is also given up. As for blocking, the endpoint executing blocking operations will be added into the queue and will hang on until it has received the connection request from the other endpoint.

5. Read/Write

When the connection between endpoints is built, the requesting end can transport data to the listening end. In general, scif_send() can be used to send data while scif_recv() is used to receive the data.

We have already introduced the messaging process between two nodes and the API functions working on such communications (see Sect. 5.5.4 for detail). Normally, the endpoint for sending requests is the local endpoint, while the one receiving requests is named the remote endpoint. The two endpoints for communication are on two different nodes. In fact, SCIF also supports connection between two endpoints within the same node, which is called a "loopback connection".

One thread can generate any number of connections; this is only limited by memory size. Figure 5.9 shows the connections between different nodes in the SCIF network. We can see that there exists a connection between node N_0 and node N_2 and also between N_1 and N_2. The connection on node N_1 is a loopback connection. In fact, in the first situation, the communication works on the host and MIC, or just between different MICs. In the second situation, the loopback connection works on different ports of one MIC, or just the CPU itself.

Fig. 5.9 Connections
between different nodes in the
SCIF network

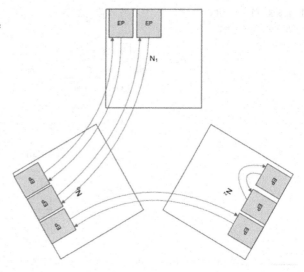

5.5.4   SCIF's API Functions

The most used API functions are listed below:

1.   scif_open()
2.   scif_bind(scif_epd_t epd, uint16_t pn)
3.   scif_listen(scif_epd_t epd, int backlog)
4.   scif_connect(scif_epd_t epd, struct scif_portID *dst)
5.   scif_accept(scif_epd_t epd, struct scif_portID *peer, scif_epd_t *newepd, int flags)
6.   scif_close()

1. scif_open(): this function is used to generate an endpoint, which will connect to
   another endpoint.
2. scif_bind(scif_epd_t epd, uint16_t pn): calling this API function can bind the
   generated endpoint to the related port. The parameter pn is the port ID.
3. scif_listen(scif_epd_t epd, int backlog): this API function assigns the endpoint
   epd to the listening one, and it can only receive a connection request from the
   other endpoint. The parameter defines the max awaiting connection number.
4. scif_connect(scif_epd_t epd, struct scif_portID *dst): this API is used to send a
   connection to the remote node. The parameter epd is the endpoint's address,
   which is returned by scif_open(). The parameter dst is the remote node's address
   information, which covers the remote node and port IDs.
5. scif_accept(scif_epd_t epd, struct scif_portID *peer, scir_epd_t *newepd, int
   flags): this API is used to receive connection requests from the remote node. The
   parameter epd denotes the returned endpoint's address from scif_open(), peer
   includes the peer node's and port's information, newepd denotes a new endpoint

that is sent back for reference in the future, and flags will denote if the mode is synchronous.

6. scif_close(): close the endpoint generated by scif_open().

Messaging Layer

Next, we introduce the API functions for messaging.

1. scif_send(scif_epd_t epd, void *msg, size_t len, int flags)

2. scif_recv(scif_epd_t epd, void *msg, size_t len, int flags)

1. scif_send(scif_epd_t epd, void *msg, size_t len, int flags): this is used to send data to the remote node. epd is the returned endpoint address from scif_open(), msg is the address for storing data, len is the length of the data transported to the remote node, and flags is for blocking or nonblocking mode. With blocking mode, it won't be sent back until all the information is sent to the remote node; otherwise, with nonblocking mode, it can be sent back directly, regardless of success.

2. scif_recv(scif_epd_t epd, void *msg, size_t len, int flags) is used to receive data from the remote node. All the parameters are the same as those of scif_open().

Remote Memory Access

API functions related to RMA:

1. off_t scif_register(scif_epd_t epd, void *addr, size_t len, off_t offset, int prot_flags, int map_flags);

2. int scif_unregister(scif_epd_t epd, off_t offset, size_t len);

1. off_t scif_register(scif_epd, void *addr, size_t len, off_t offset, int prot_flags, int map_flags): this is used to register its own buffers for remote memory access, the address of which starts from self_addr, with a size of msg_size.

2. int scif_unregister(scif_epd_t epd, off_t offset, size_t len): used to delete the registered window, which has the same size as scif_register(), which is [offset, offset+msg_size-1].

Remote mapping

Related API functions are listed below:

1. void *scif_mmap(void *addr, size_t len, int prot, int flags, scif_epd_t epd, off_t offset);

2. int scif_munmap(void *addr, size_t len);

1. void *scif_mmap(void *addr, size_t len, int prot, int flags, scif_epd_t epd, off_t offset): used to map the virtual page address to the remote window. All the pages of this window are mapped with all the physical pages. The address of this window starts at addr, with a size of len, and the related physical pages denote the registered pages by new epd.

2. int scif_mummap(void *addr, size_t len): this function is used to remove the mapping generated by the scif_mmap() functions.

RMA Transportation

On the local physical memory and the remote node of special node, scif_readfrom() and scif_writeto() functions are executed to read and write based on the DMA or CPU.

scif_vreadfrom() and scif_vwriteto() functions are the alternatives of scif_readfrom () and scif_writeto(). When the local range is the only source or the transporting destination, using scif_vreadfrom() and scif_vwriteto() can improve performance.

1.   int scif_readfrom(scif_epd_t epd, off_t loffset, size_t len, off_t roffset, int rma_flags);

2.   int scif_writeto(scif_epd_t epd, off_t loffset, size_t len, off_t roffset, int rma_flags);

3.   int scif_vreadfrom(scif_epd_t epd, off_t *addr, size_t len, off_t roffset, int rma_flags);

4.   int scif_vreadfrom(scif_epd_t epd, off_t *addr, size_t len, off_t roffset, int rma_flags);

1. int scif_readfrom(scif_epd_t epd, off_t loffset, size_t len, off_t roffset, int rma_flags): this function is used to read data from the remote node. The data size of the remote registered address is msg_size, and the address starts at buffers [i].offset. The data is copied to the local registered address, which starts at buffers[i].offset.
2. int scif_writeto(scif_epd_t epd, off_t loffset, size_t len, off_t roffset, int rma_flags): this function is used to write data to the remote node. The data with a size of msg_size is copied from the local registered address to the remote registered address. The initial address is the same as scif_readfrom.
3. int scif_vreadfrom(scif_epd_t epd, off_t *addr, size_t len ,off_t roffset, int rma_flags): this function is an alternative of scif_readfrom. It's also used to read remote data to the local address. The difference is that the address of scif_readfrom is the physical address mapped from the local registered address while scif_vreadfrom is mapped from the local virtual address.
4. int scif_vwriteto(scif_epd_t epd, off_t *addr, size_t len, off_t roffset, int rma_flags): this function is an alternative to scif_writeto, with the same method of writing data to the remote address from the local address. The difference between scif_writento and scif_vwriteto is similar to the difference between scif_readform and scif_vreadfrom, discussed above.

Synchronization of SCIF

Synchronization of SCIF works for RMA. The related API functions are listed below:

1.   int scif_fence_mark(scif_epd_t epd, int flags, int64_t *mark);

2.   int scif_fence_wait(scif_epd_t epd, int64_t mark);

1. int scif_fence_mark(scif_epd_t epd, int flags, int64_t *mark): this function is used to mark the initialized RMA operations, making them synchronous.
2. int scif_fence_wait(scif_epd_t epd, int64_t mark): this function is used to make all the marked RMA operations wait.

Now, we have finished introducing the foundation of SCIF communication. For detailed examples and optimization, please see Chap. 8.

Debugging and Profiling Tools for the MIC

6

Intel has lots of experience in integrating software and hardware systems to support high-performance foundational software such as the compiler, debugger, math library, profiler, and others. Under this tradition, MIC is fully supported by Intel's parallel toolkits. This chapter focuses on the unique methods for the MIC platform that differ from the common ones, since MIC is compatible with the traditional x86 architecture. The Intel debugger (IDB) tool and the profiling tool Intel VTune Amplifier XE are introduced later in the chapter.

Chapter Objectives.

- to learn the supported Intel software toolkits for MIC
- to learn the debugging tool IDB for MIC
- to learn the profiling tool Intel VTune Amplifier XE for MIC

6.1 Intel's MIC-Supported Tool Chains

MIC supports programming based on pragmas, for which some MIC-related precompiled instructions are added to the codes. With an extended MIC coprocessor, the related codes will run on the MIC device; otherwise, they will run on the host CPU device.

MIC-supported components in Intel Composer XE 2013 are listed below:

- Intel C++/Fortran compilers
- Intel debugger (IDB)
- Intel Math Kernel Library (MKL)
- Intel Threading Building Blocks (TBB)
- Eclipse IDE integration

Intel's profiling tool—Intel VTune Amplifier XE—also fully supports MIC, which can improve optimizing efficiency.

E. Wang et al., *High-Performance Computing on the Intel® Xeon Phi™*,
DOI 10.1007/978-3-319-06486-4_6, © Springer International Publishing Switzerland 2014

6.2   MIC Debugging Tool IDB

6.2.1   Overview of IDB

IDB is a debugging tool from Intel used with symbol-based codes. It can perform the following operations:

1. debug C/C++/Fortran codes
2. disassemble or check machine codes, check the register value
3. debug multi-thread codes (only for Linux OS)
4. debug MIC codes (only for Linux OS)

IDB supports debugging C++ based parallel codes under Linux OS. It extends Intel C++ Compiler, Intel Cilk Plus, and OpenMP runtime for debugging codes. IDB's parallel debugging features include:

1. analysis of the threading shared data, which checks the contemporary access of different threads (C/C++/Fortran)
2. intelligent breakpoints when the interrupt function is called repeatedly by different threads
3. checking the vector register, such as the Intel SSE extended instruction registers, to make IDB debug SIMD instruction-level parallel codes
4. simulating OpenMP or Intel Cilk Plus in serial execution
5. profiling OpenMP runtime information

6.2.2   IDB Interface

IDB has two debugging interfaces: the graphical user interface (GUI) and the command-line based interface. Both interfaces support debugging codes for the MIC platform.

1. GUI debugging

   To start the GUI debugging interface, run idb_mpm in <install_dir>/bin/ intel64_mic. The user can then track the codes running on MIC and debug them. For remote debugging, one needs to use SSH with the environmental variable DISPLAY to login. Using the GUI, IDB can fully control the debugging process. Most of the functions, such as single-step, step-through-function, and memory print, can be launched by just clicking the button on the toolbar.

2. Command-line debugging

   There are two sub-versions for IDB command-line debugging: idbc for IA64 and idbc_mic for MIC. One thing to note is that there is no autotracking function for command-line debugging.

6.2.3 IDB Support and Requirements for MIC

IDB supports MIC with new features, such as:

1. debugging MIC codes for offload mode and native mode
   a. for offload mode, idb GUI (idb) or idb Command (idbc) can both be used
   b. for native mode, only idbc_mic can be used
2. showing OpenMP objects because OpenMP is the most common parallel model for MIC multi-threading
3. controlling a specified thread
4. checking and assigning a value for the 512-bit registers on the MIC VPU
5. real-time calculation of the expressions for debugging.

There are also some precautions to consider when debugging MIC codes with IDB:

1. When command-line debugging with native mode, one needs to upload idbserver_mic onto the MIC device and then start it by logging in on the MIC card through SSH. This also means that using idbc_mic should make the user ID the same as the one on MIC. Otherwise, the MIC device must be set to no-password login for SSH.
2. When command-line debugging with the native mode, all the dynamic libraries need to be uploaded onto the MIC device.
3. When command-line debugging for heterogeneous programs, the offload process is actually launched by the root. Without the root account, idbc_mic may have the problem of idbserver_mic not being able to access the offload process. One method to overcome this is to start idbc_mic with root authority. Some other options can also be added when starting idbc_mic: -mpm-launch –mpm-cardid=<card-id>:

```
idbc_mic -mpm-launch= 1 -mpm-cardid=<card -id> -tco -rconnect=tcpip:<cardip>:<port>
```

4. When IDB GUI debugging for heterogeneous programs, there may be errors such as "offload error: cannot start processes on the device 0 (error code 1)" when generating offload processes. One can overcome this by restarting the debugging session until success is achieved.

6.2.4 Debugging MIC Programs Using IDB

Follow the steps below to debug MIC programs with IDB:

1. Preparation: setting environment, compiling with debugging information, linking dynamic libraries, etc.
2. Setting debugger: configuring the debugging object, setting parameters, setting breakpoints.
3. Running the program: the program stops at breakpoints so the user can debug it with single-step execution.

4. Monitoring codes, data, and process information: checking the memory, register, variable, thread/process ID, etc.
5. Locating errors.
6. Correcting errors.
7. Repeating the steps above.

6.2.4.1 Preparation

Configure the debugging environment using PATH, LD_LIBRARY_PATH, SHELL and HOME.

1. Configure the environmental variables (only for gdb)
   Use the following command:
   set environment *name* [*value*]
   in which:
   name is the name of variable.
   value is the value of variable.
   The following example shows how to configure environmental variables for IDB:

```
(idb) show environment FOO
Environment variable "FOO" not defined.
(idb) set environment FOO = rabbit
set environment FOO = rabbit
                    ^
Unable to parse input as legal command or C expression.
(idb) set environment FOO=rabbit
(idb) show environment FOO
FOO==rabbit
(idb)set environment FOO rabbit
(idb) show environment FOO
FOO=rabbit
(idb)
```

2. Check the environmental variables
   Use the following command to check the environmental variables:
   show environment [*name*]
   in which *name* is the name of environmental variable.
   The following example shows how to check environmental variables for IDB:

```
(idb) show environment USER
USER=hal
```

6.2.4.2 Compile with Debugging Information

To add debugging information for the binary program, one needs to add the –g option when compiling codes.

```
[root@mic5 matrixMul]# make mic
icc -g -O3 -vec-report -D__USE_OMP__ -openmp -offload-build -D__USE_MIC__  -o matrix_mic matrix.c -lm
matrix.c(97): (col. 29) remark: *MIC* LOOP WAS VECTORIZED.
matrix.c(36): (col. 13) remark: *MIC* LOOP WAS VECTORIZED.
matrix.c(36): (col. 13) remark: *MIC* LOOP WAS VECTORIZED.
matrix.c(97): (col. 29) remark: LOOP WAS VECTORIZED.
matrix.c(36): (col. 13) remark: LOOP WAS VECTORIZED.
matrix.c(36): (col. 13) remark: LOOP WAS VECTORIZED.
```

Without the debugging information, the codes cannot be related to the debugging process.

6.2.4.3 Start Debugging

Next, we introduce the methods of using IDB GUI and IDB Command, respectively, for debugging MIC programs.

Using IDB GUI to debug MIC programs

1. Start IDB GUI

   IDB is a component of Intel composer_xe, which is located in intel/composer_xe_2013/bin/. Enter this directory to start IDB:

```
[root@mic4 intel64]# idb
Xlib:  extension "RANDR" missing on display "localhost:11.0".
```

2. Open/Close console window

   Users can click the button 🖻 to open or close the console window. Debugger commands are shown in this window (Fig. 6.1).
3. Load debugging object

   IDB supports two methods for loading debugging objects: opening executable files directly and assigning a process ID.

   a. Open executable file directly: click the button File->Open Executable on the toolbar. The pop-up window is shown in Fig. 6.2:

   Choose the executable file for debugging and fill in the necessary arguments for the parameters (Fig. 6.3).

   b. Debug the assigned process. Check the process ID first.

```
[root@mic4 ~]# ps  -ef |grep matrix
root      31928 31612 53 10:40     pts/9     00:00:56 ./matrix_mic 120 102400
root      31934 31875  0 10:42 pts/8     00:00:00 grep matrix
[root@mic4 ~]#
```

Fill in the process ID in the pop-up window.
Click File->Attach To Process in the menu (Fig. 6.4).

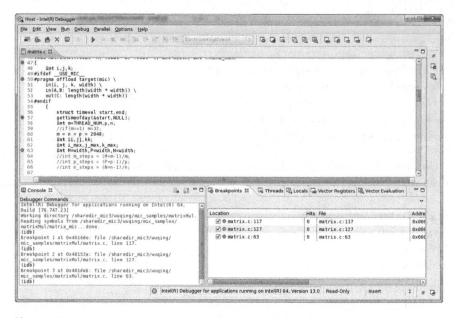

Fig. 6.1 Open/Close console window

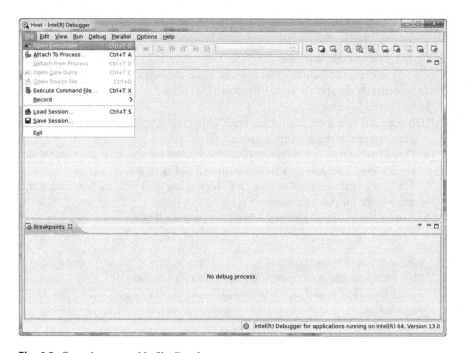

Fig. 6.2 Open the executable file directly

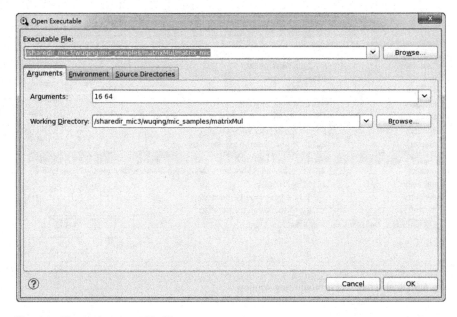

Fig. 6.3 Choose the executable file

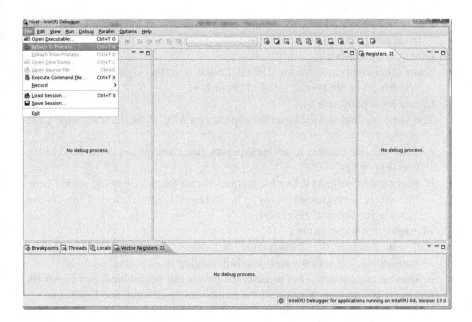

Fig. 6.4 Choose the tracking process

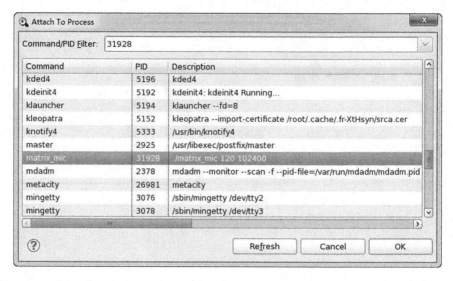

Fig. 6.5 Real-time process monitoring window

The pop-up real-time monitoring window is shown in Fig. 6.5.

Double-click the selected process and IDB will track it into the debugging process (Fig. 6.6).

4. Configure related codes

   Figure 6.7 shows adding the source code directory to let IDB check related code files.

5. Other configurations

   Some other configurations for IDB GUI, such as setting style of the debugging window, will not be discussed in detail here.

6. Create breakpoints for MIC

   The same method is used to set breakpoints on MIC as on the traditional CPU platform.

   a. Click the line number to set breakpoints. Just click the dot at the head of the line (Fig. 6.8).

   b. Right-click to create a line breakpoint. Hover the mouse at the source code line where a breakpoint needs to be added and right-click it. Then choose "Create Breakpoint" (Fig. 6.9).

7. Managing the breakpoints

   IDB can manage the breakpoints, including "close", "delete", and "set breakpoints". It also supports batch-managing breakpoints.

   a. Disable breakpoints. One or all of the points can be disabled. See Fig. 6.10.

   b. Delete breakpoints. One or all the points can be deleted. See Fig. 6.11.

   c. Set the breakpoints' properties. The properties, such as triggering conditions, breakpoint behaviors, ignoring numbers, and process filters, can be assigned. See Fig. 6.12.

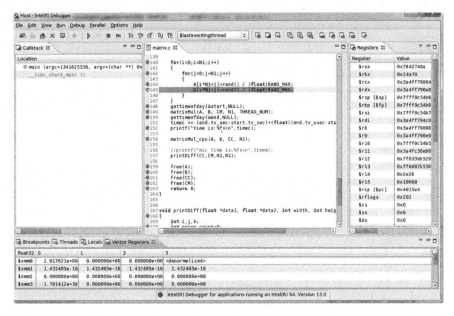

Fig. 6.6 Track into the debugging process

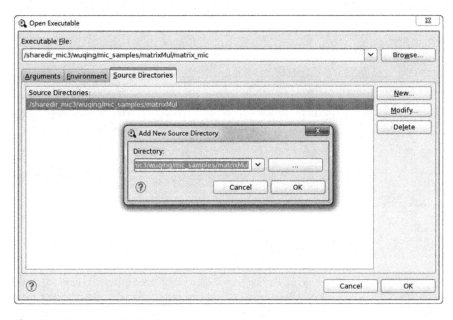

Fig. 6.7 Add source code directory

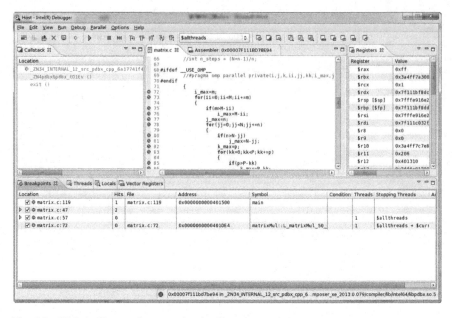

Fig. 6.8 Click the line number to set breakpoints

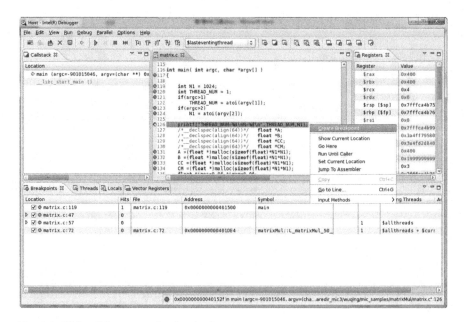

Fig. 6.9 Right-click to create line breakpoint

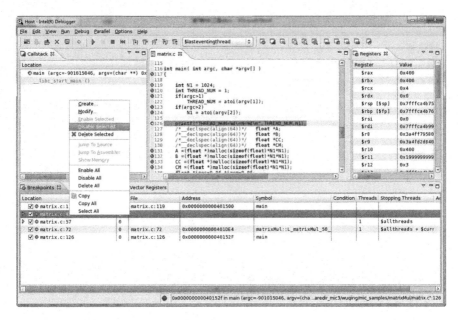

Fig. 6.10 Disable breakpoints

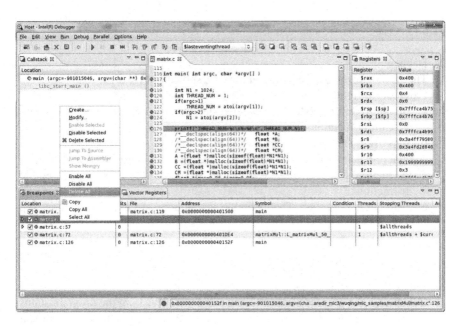

Fig. 6.11 Delete breakpoints

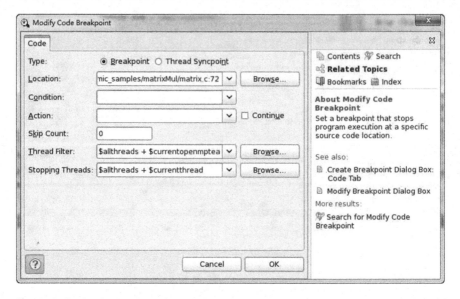

Fig. 6.12 Set breakpoint properties

Fig. 6.13 Control buttons for the IDB debugger

8. Debugging
   The IDB debugger's executing buttons are shown in Fig. 6.13.
   From left to right:
   a. Step into: Source codes are executed step by step. Execution will track into the calling functions.
   b. Step over: The functions are executed as one step.
   c. Function debug: Execute from the current location until function is over.
   d. Instruction step into: Source codes are executed step-by-step as assemble instructions.
   e. Instruction step over: The assemble instructions are executed as one step (Fig. 6.14).
   f. Jump to the specified location and stop.
      Right-click and choose "Go Here", and the program will stop at the specified location (Fig. 6.15).
9. Check the program status
   Debugging is a process of discovering problems. Checking the program status can dig and locate the program's bugs. IDB GUI can check the program status in real-time; the status bar is shown in Fig. 6.16.
   The functions of the buttons on the status bar listed below.
   a. View the breakpoints (Fig. 6.17):
   b. View the stack information (Fig. 6.18):

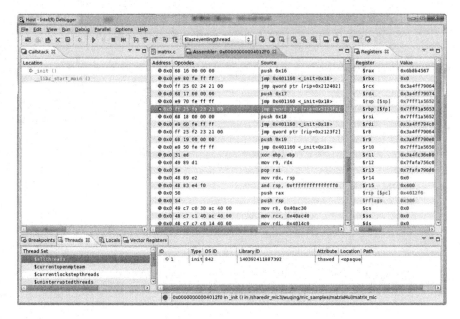

Fig. 6.14 Instruction step over

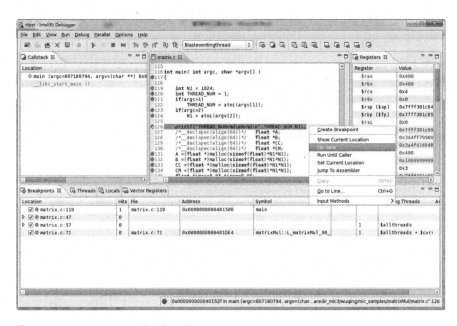

Fig. 6.15 Jump to the specified location

Fig. 6.16 Program status bar

Location	Hits	File	Address	Symbol	Condition	Threads	Stopping Threa
☑ ◉ matrix.c:119	0	matrix.c:119	0x0000000000401500	main			
▷ ☑ ◉ matrix.c:47	0						
▷ ☑ ◉ matrix.c:57	0					1	$allthreads
☑ ◉ matrix.c:72	0	matrix.c:72	0x0000000000401DE4	matrixMul::L_matrixMul_50_		1	$allthreads

Fig. 6.17 View breakpoints

Location	File	Module	Address
◇ _ZN34_INTERNAL_12_src_pdbx_cpp_6a17741f4pdbx21o171		libpdbx.so.5	0x00007FB6CDAFDE94
_ZN4pdbx6pdbx_tDIEv ()		libpdbx.so.5	0x00007FB6CDAFF6DA
exit ()		libc-2.12.so	0x0000003A4FC35FD2
_ZN3MIC6Engine4initE11PPKc ()		liboffload.so.5	0x00007FB6D027C F79
_ZN3MIC4initEv ()		liboffload.so.5	0x00007FB6D027D8E0
__offload_init ()		liboffload.so.5	0x00007FB6D02847D2

Fig. 6.18 View the stack information

Thread Set	ID	Type	OS ID	Library ID	Attribute	Location	Path
$allthreads	1	init	2229(	140695132841760	thawed	void ma	/sharedir_mic3/wuc
$currenttopenmpteam	2	unkn	2229!	140695159453456	thawed	<opaque	
$currentlockstepthreads	3	omp	2229:	140695100487440	thawed	void ma	/sharedir_mic3/wuc
$uninterruptedthreads	4	omp	2229(	140695096289040	thawed	void ma	/sharedir_mic3/wuc
$frozenthreads	◇ 5	omp	2229/	140695092090640	thawed	void ma	/sharedir_mic3/wuc
◇ $lasteventingthread	6	omp	2229(	140695087092240	thawed	void ma	/sharedir_mic3/wuc
$currentthread	7	omp	2229(	140695083693840	thawed	void ma	/sharedir_mic3/wuc
	8	omp	2229:	140694740686672	thawed	void ma	/sharedir_mic3/wuc

● 0x000000000401a6b in matrixMul (A=(float *) 0x0, B=(floa...'/sharedir_mic3/wuqing/mic_samples/matrixMul/matrix.c':91

Fig. 6.19 View process information/Switch process

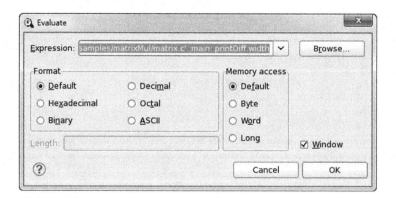

Fig. 6.20 Add Evaluate

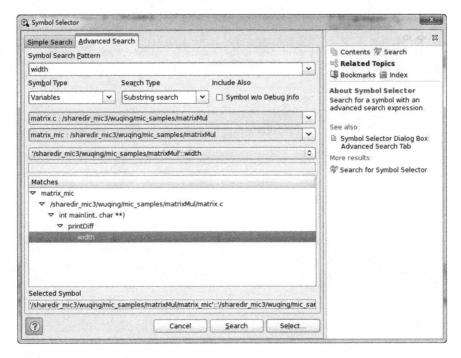

Fig. 6.21 Search variables/expression symbols

   c. View process information/Switch process (Fig. 6.19).

      Double-clicking any process can switch to the selected process for debugging.

   d. View/Set expression values. In IDB's Evaluate window, "Expression" can be set as symbols, expressions, variables, and arrays (Fig. 6.20).

      Users can search variables/expression symbols using keywords (Fig. 6.21).

      The value of Expression can be viewed, modified, or formatted in the run time (Fig. 6.22).

      See Fig. 6.23 on how to view the value of local variables.

   e. View the status of user-defined vectorized expression.

   f. View the information of assembly codes (Fig. 6.24).

   g. View the common register (Fig. 6.25).

   h. View the memory section. In this example, we view the values of matrix A. See Figs. 6.26 and 6.27.

   i. View the VPU vectorized register of MIC core. Each VPU of MIC core has 16 registers (xmm0~xmm15), and each register can contain 4 float32 (Fig. 6.28).

   j. Open/Close the source code window. IDB can display the source code tree (Fig. 6.29).

10. Correct the code and repeat the steps above.

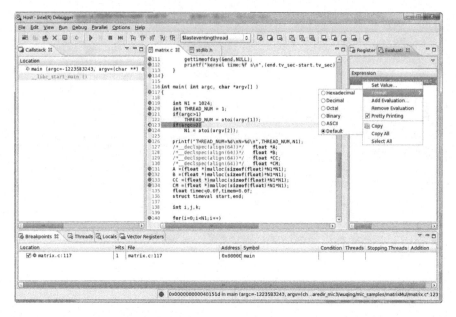

Fig. 6.22 Modify the value of Expression

Expression	Value	Type
▷ A	0x1d09010	float *
▷ B	0x1d0d020	float *
▷ C	0x1d15040	float *
THREAD_NUM	16	int
i	-1409288172	int
j	32767	int
k	-1409422720	int
width	64	int

Fig. 6.23 View the value of local variables

According to the problems found, we can correct and improve the code, and then test and debug the code over again.

Using IDB Command to debug MIC codes

There are two debugging modes: host debugger and target debugger.

1. Host debugger

This is used to debug programs running on the host device and is suitable for running programs on the host of computing applications that use both the CPU and MIC. The MIC part is executed with offload mode. The debugger idbc is located at /opt/intel/composer_xe_2013.0.079/bin/intel64. The commands are:

```
[root@mic4 matrixMul]# idbc
Intel(R) Debugger for applications running on Intel(R) 64, Version 13.0, Build [78.747.23]
(idb)
```

Address	Opcodes	Source
0x0	8b 74 24 28	mov esi, dword ptr [rsp+0x28]
0x0	46 8d 7c 00 ff	lea r15d, ptr [rax+r8*1-0x1]
0x0	41 0f af ef	imul ebp, r15d
0x0	48 8b 7c 24 08	mov rdi, qword ptr [rsp+0x8]
0x0	42 8d 0c 2e	lea ecx, ptr [rsi+r13*1]
0x0	03 e9	add ebp, ecx
0x0	8d 76 ff	lea esi, ptr [rsi-0x1]
0x0	48 63 ed	movsxd rbp, ebp
0x0	44 8b 6c 24 18	mov r13d, dword ptr [rsp+0x18]
0x0	48 8b 4c 24 30	mov rcx, qword ptr [rsp+0x30]
mat		for(j=jj; j<jj+j_max; j++)
0x0	89 44 24 40	mov dword ptr [rsp+0x40], eax
0x0	89 5c 24 38	mov dword ptr [rsp+0x38], ebx
0x0	4c 8d 04 af	lea r8, ptr [rdi+rbp*4]
0x0	8b 7c 24 70	mov edi, dword ptr [rsp+0x70]
0x0	44 0f af ff	imul r15d, edi
0x0	0f af f7	imul esi, edi
0x0	48 8b 44 24 60	mov rax, qword ptr [rsp+0x60]
0x0	44 03 ef	add r13d, edi
0x0	45 03 fd	add r15d, r13d
0x0	44 03 ee	add r13d, esi
0x0	4d 63 ff	movsxd r15, r15d
0x0	4e 8d 3c b9	lea r15, ptr [rcx+r15*4]
0x0	44 89 e1	mov ecx, r12d
0x0	4c 89 fd	mov rbp, r15
0x0	48 83 e5 0f	and rbp, 0xf
0x0	89 ee	mov esi, ebp
0x0	89 ef	mov edi, ebp
0x0	f7 de	neg esi

● 0x0000000000401b8e in matrixMul (A=<no value...3/wuqing/mic_samples/matrixMul/matrix.c':100

Fig. 6.24 View the information of assembly codes

Register	Value	Description
$rax	0x9dde4f9	165537017
$rbx	0xafd80e58	2950172248
$rcx	0x400bf7	4197367
$rdx	0x0	0
$rsp [$sp]	0x7fffafd80c90	(void *) 0x7fffafd80c90
$rbp [$fp]	0xffffffff8277fbff	(void *) 0xffffffff8277fbff
$rsi	0xffffffff	4294967295
$rdi	0x7f82098bc000	140196482629632
$r8	0x86ed940	141482304
$r9	0x0	0
$r10	0x0	0
$r11	0x0	0
$r12	0x7f8207898010	140196448927760
$r13	0x400bf7	4197367
$r14	0x86ed940	141482304
$r15	0x124cbe38	307019320
$rip [$pc]	0x401b8e	(void *) 0x401b8e
$rflags	0x286	646
$cs	0x0	0
$ss	0x0	0
$ds	0x0	0
$es	0x0	0
$fs	0x0	0
$gs	0x0	0
$fctrl	0x37f	895
$fstat	0x0	0
$ftag	0x0	0
$fop	0x0	0
$fiseg	0x0	

● 0x0000000000401b8e in matrixMul (A=<no value...3/wuqing/mic_samples/matrixMul/matrix.c':100

Fig. 6.25 View showing the common registers

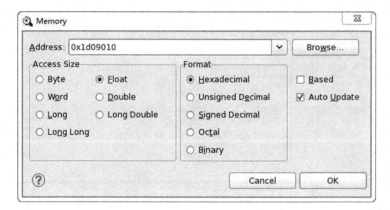

Fig. 6.26 View showing the memory section

Fig. 6.27 View memory section

2. Target debugger
   Used to debug programs running on MIC normally based on two conditions:

- MIC programs running with offload mode. The CPU/MIC mixed computing application runs the MIC part with offload mode.
- MIC programs running with native mode. The debugger for the target debugger mode is idbc_mic. The IDB Command (idbc) debugger cannot track with MIC

float32	0	1	2	3
$xmm0	0.000000e+00	0.000000e+00	0.000000e+00	0.000000e+00
$xmm1	0.000000e+00	0.000000e+00	0.000000e+00	0.000000e+00
$xmm2	0.000000e+00	0.000000e+00	0.000000e+00	0.000000e+00
$xmm3	1.300338e-08	4.074358e-11	1.050117e-05	1.709001e+25
$xmm4	1.804184e+28	3.522579e-06	5.423024e+28	1.325699e+25
$xmm5	1.109575e+27	4.630426e+27	7.184942e+22	7.330672e+22
$xmm6	2.564000e-09	4.005880e-11	1.684526e-10	2.168341e-10
$xmm7	7.214338e+22	1.298569e+19	8.675072e-04	7.619667e+31
$xmm8	7.555542e+31	3.235991e+21	7.440870e+28	1.746649e+19
$xmm9	1.300338e-08	4.074358e-11	1.050117e-05	6.734817e+22
$xmm10	0.000000e+00	0.000000e+00	0.000000e+00	0.000000e+00
$xmm11	0.000000e+00	0.000000e+00	0.000000e+00	0.000000e+00
$xmm12	0.000000e+00	0.000000e+00	0.000000e+00	0.000000e+00
$xmm13	0.000000e+00	0.000000e+00	0.000000e+00	0.000000e+00
$xmm14	0.000000e+00	0.000000e+00	0.000000e+00	0.000000e+00
$xmm15	0.000000e+00	0.000000e+00	0.000000e+00	0.000000e+00

Fig. 6.28 View VPU vectorized register of MIC core

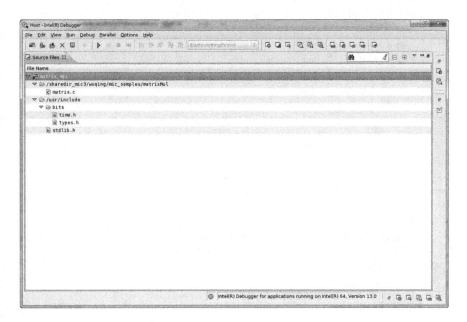

Fig. 6.29 Source code tree

programs, so for CPU/MIC mixed computing applications, we suggest using Eclipse-based GUI idb to debug both host and target programs.

Next, we introduce the two situations mentioned above for IDB Command.
1. Debug the MIC program with offload mode.
   Step 1: Preparation
(a) Modify codes. Because idbc cannot track MIC programs, we need to add an infinite loop manually at the head of the MIC program:

```
[pseudo-code]
......
__declspec(target(mic)) void attach_idb() {
    volatile int loop = 1;
    do {
        volatile int a = 1;
    } while (loop);
}
......
void matrixMul_MIC(float *A, float *B, float *C, int width, int THREAD_NUM)
{
        int i,j,k;
#ifdef __USE_MIC__
#pragma offload target(mic)    \
        in(i, j, k, width)  \
        in(A,B: length(width * width))  \
        out(C:  length(width * width))
#endif
        {
                attach_idb();
                ......
        }
}
......
int main( int argc, char *argv[] )
{
        ......
        matrixMul_MIC(A, B, C, width, THREAD_NUM);
        ......
}
```

(b) Compile codes with the –g option.

```
[root@mic4 matrixMul]# make mic
icc -g -O3 -vec-report -D__USE_OMP__ -openmp -offload-build -D__USE_MIC__    -o matrix_mic matrix.c -lm
matrix.c(43): (col. 5) remark: *MIC* LOOP WAS VECTORIZED.
matrix.c(43): (col. 5) remark: *MIC* LOOP WAS VECTORIZED.
matrix.c(105): (col. 9) remark: *MIC* LOOP WAS VECTORIZED.
matrix.c(43): (col. 5) remark: LOOP WAS VECTORIZED.
matrix.c(43): (col. 5) remark: LOOP WAS VECTORIZED.
matrix.c(105): (col. 9) remark: LOOP WAS VECTORIZED.
[root@mic4 matrixMul]#
```

If the program runs with native mode on MIC by itself, we should definitely employ the –mmic option:

```
[root@mic4 matrixMul]# make native
icc -g -O3 -vec-report -D__USE_OMP__ -openmp -mmic -D__USE_MIC__    -o matrix_native matrix.c -lm
matrix.c(105): (col. 9) remark: LOOP WAS VECTORIZED.
matrix.c(43): (col. 5) remark: LOOP WAS VECTORIZED.
matrix.c(43): (col. 5) remark: LOOP WAS VECTORIZED.
matrix.c(43): (col. 5) remark: LOOP WAS VECTORIZED.
matrix.c(105): (col. 9) remark: LOOP WAS VECTORIZED.
[root@mic4 matrixMul]#
```

Step 2: Start idbc_mic

The idbc_mic is located at /opt/intel/composer_xe_2013.0.079/bin/intel64_mic. The starting command is:

```
idbc_mic  -tco -rconnect=tcpip:coprocessor    -ip-address:port
```

For example:

```
[root@mic4 matrixMul]# idbc_mic -tco -rconnect=tcpip:172.31.1.1:2000
idbserver_mic 100% 1609KB    1.6MB/s    00:00
Intel(R) Many Integrated Core Architecture Debugger (Intel(R) MIC Debugger) Version 13.0, Build [78.833.23]
(idb)
```

Step 3: Set PATH

The user can set PATH to find the library. The command is:

```
set solib-search-path path[:path]
```

For example:

```
(idb) set solib -search -path /tmp
(idb) set solib -search -path /lib64
(idb) set solib -search -path /sharedir_mic3/wuqing/mic_samples/matrixMul/
```

Step 4: Set environmental variables on MIC

If the MIC program cannot find the library while debugging, one can upload the library to MIC using the scp command, and SSH to set MIC-related environmental variables. For example:

```
[root@mic4 lib]# scp * mic0:/lib64
libmyo -client.so    100%   216KB 216.2KB/s    00:00
libmyodbl -client.a  100%    21KB  20.7KB/s    00:00
libmyodbl -client.so 100%    17KB  17.2KB/s    00:00
libmyodbl -service.a 100%    22KB  21.6KB/s    00:00
libmyodbl -service.so 100%   18KB  17.8KB/s    00:00
libmyo -service.so   100%   245KB 245.4KB/s    00:00
[root@mic4 lib]# SSH  mic0
[root@mic4-mic0 /tmp]# export LD_LIBRARY_PATH=/tmp
[root@mic4-mic0 /tmp]# echo $LD_LIBRARY_PATH
/tmp
[root@mic4-mic0 /tmp]#
```

Step 5: Start application
Start CPU/MIC mixed application on the host:

```
[root@mic4 matrixMul]# ./matrix_mic 16 64
THREAD_NUM=16
N=64
```

Now the MIC program is running the infinite loop which we added in the beginning.

Step 6: Catch and track the offload_main process on MIC

First, we log into the MIC device through SSH. We can write the /etc/hosts to add the IP address for MIC.

```
[root@mic4 ~]# SSH mic0
[root@mic4-mic0 /root]# cd /tmp
[root@mic4-mic0 /tmp]# ls
coi_procs     idbserver_mic  libiomp5.so    matrix.c      matrix_native
[root@mic4-mic0 /tmp]#
[root@mic4-mic0 /tmp]# ifconfig
lo        Link encap:Local Loopback
          inet addr:127.0.0.1   Mask:255.0.0.0
          inet6 addr: ::1/128 Scope:Host
          UP LOOPBACK RUNNING    MTU:16436  Metric:1
          RX packets:0 errors:0 dropped:0 overruns:0 frame:0
          TX packets:0 errors:0 dropped:0 overruns:0 carrier:0
          collisions:0 txqueuelen:0
          RX bytes:0 (0.0 B)   TX bytes:0 (0.0 B)
```

```
mic0        Link encap:Ethernet   HWaddr 26:85:19:0A:B5:C7
            inet addr:172.31.1.1   Bcast:0.0.0.0   Mask:255.255.255.0
            inet6 addr: fe80::5c78:a2ff:fec9:ef04/64 Scope:Link
            UP BROADCAST RUNNING    MTU:65535   Metric:1
            RX packets:7377 errors:0 dropped:0 overruns:0 frame:0
            TX packets:7637 errors:0 dropped:0 overruns:0 carrier:0
            collisions:0 txqueuelen:1000
            RX bytes:23154514 (22.0 MiB)    TX bytes:17463577 (16.6 MiB)
[root@mic4-mic0 /tmp]#
```

Then we can view the process ID of offload_main on MIC using the ps command:

```
[root@mic4-mic0 /tmp]# ps -ef | grep offload
19178 root            0:17 /tmp/coi_procs/1/19178/offload_main
19184 root            0:00 grep offload
[root@mic4-mic0 /tmp]#
```

The offload_main process on MIC can be tracked with "attach", using the following command, in which pid is the process ID of offload_main on MIC that we obtained in the last step.

```
attach    pid   /opt/intel/composerxe_mic/compiler/lib/mic/offload_main
```

An example is given below:

```
(idb) attach 19178 /opt/intel/composer_xe_2013.0.079/compiler/lib/mic/offload_main
Attaching to program: /opt/intel/composer_xe_2013.0.079/compiler/lib/mic/offload_main, process 19178
[New Thread 19178 (LWP 19178)]
Reading symbols from /opt/intel/composer_xe_2013.0.079/compiler/lib/mic/offload_main...done.
[New Thread 19180 (LWP 19180)]
[New Thread 19181 (LWP 19181)]
[New Thread 19182 (LWP 19182)]
_pthread_cleanup_pop_restore () in /lib64/libpthread-2.12.so
 (idb) h
List of classes of commands:
      breakpoints -- Commands for manipulation with breakpoints.
      data        -- Examining data.
      extensions  -- Idb extension commands.
      files       -- Specifying and examing files.
      obscure     -- Obscure features.
      openmp      -- OpenMP support.
      parallel    -- MPI support.
      running     -- Running the program.
      stack       -- Examining the stack.
      status      -- Status inquiries.
      support     -- Support facilities.
To display help on a particular command, enter "help" followed by the command
name. Command name abbreviations are allowed if unambiguous.
(idb)
```

Step 7: Set breakpoints

There is no big difference between breakpoint manager idbc_mic and traditional idb. We can set breakpoints through line number, function symbol, etc. Use info breakpoints to view the breakpoint information:

```
(idb) b 55
Breakpoint 2.1 at 0x400dc2: file /sharedir_mic3/wuqing/mic_samples/matrixMul/matrix.c, line 65.
Breakpoint 2.2 at 0x405a6d: file /sharedir_mic3/wuqing/mic_samples/matrixMul/matrix.c, line 65.
(idb) b main
Breakpoint 3 at 0x400b40: file /sharedir_mic3/wuqing/mic_samples/matrixMul/matrix.c, line 125.
(idb) info breakpoints
Num      Type            Disp Enb Address              What
1        breakpoint      keep y   0x0000000000400b40 in main at
/sharedir_mic3/wuqing/mic_samples/matrixMul/matrix.c:125
2        breakpoint      keep y   <MULTIPLE>
/sharedir_mic3/wuqing/mic_samples/matrixMul/matrix.c:55
2.1                      keep y   0x0000000000400dc2 in matrixMul at
/sharedir_mic3/wuqing/mic_samples/matrixMul/matrix.c:65
2.2                      keep y   0x0000000000405a6d in
matrixMul_cpu::L_matrixMul_cpu_33__par_loop0_2_193 at
/sharedir_mic3/wuqing/mic_samples/matrixMul/matrix.c:65
3        breakpoint      keep y   0x0000000000400b40 in main at
/sharedir_mic3/wuqing/mic_samples/matrixMul/matrix.c:125
(idb)
```

Step 8: Modify the loop value to jump out of the infinite loop

We must make the program jump out of the infinite loop to debug the user code section. The loop condition can be modified in the following way:

```
(idb) print loop
$2 = 1
(idb) set loop = 0
(idb) print loop
$3 = 0
(idb)
```

Step 9: Debug MIC programs

Use the continue command to continue executing until the breakpoint is reached:

```
(idb) c
Continuing.
Breakpoint 2.1, matrixMul (A=0x7fffc67783ff, B=0xff46d92d00000010, C=0x40, width=Info: symbol width is
defined but not allocated (optimized away)
<no value>, THREAD_NUM=1) at /sharedir_mic3/wuqing/mic_samples/matrixMul/matrix.c:65
65              gettimeofday(&start,NULL);
(idb)
```

Now, the actual MIC program debugging starts. The process is similar to debugging with the traditional multi-thread CPU program. Please refer to the Intel IDB manual for details.

1. Debug MIC program with native mode
   Step 1: Preparation
a. Modify codes. We also need to manually add a section of infinite loop because idbc cannot track with the MIC program. Please see the section pertaining to MIC program debugging with offload mode.
b. Compile the source codes. The program runs on MIC with the native mode. The options –g and –mmic should be added when compiling.

```
[root@mic4 matrixMul]# make native
icc -g -O3 -vec-report -D__USE_OMP__ -openmp -mmic -D__USE_MIC__   -o matrix_native matrix.c -lm
matrix.c(105): (col. 9) remark: LOOP WAS VECTORIZED.
matrix.c(43): (col. 5) remark: LOOP WAS VECTORIZED.
matrix.c(43): (col. 5) remark: LOOP WAS VECTORIZED.
matrix.c(43): (col. 5) remark: LOOP WAS VECTORIZED.
matrix.c(105): (col. 9) remark: LOOP WAS VECTORIZED.
[root@mic4 matrixMul]#
```

Step 2: Upload the executable program and library onto MIC
Upload the compiled executable program and linking library onto the MIC device through scp.

```
[root@mic4 matrixMul]# scp matrix_native mic0:/tmp
matrix_native   100%    48KB  47.8KB/s    00:00
[root@mic4 mic]# scp libiomp5.so mic0:/tmp
libiomp5.so   100%   956KB 955.6KB/s    00:00
[root@mic4 mic]#
```

Step 3: Set environmental variables (optional)
If necessary, system environmental variables such as LD_LIBRARY_PATH should be set to ensure that the MIC program can be started correctly.

```
[root@mic4-mic0 /tmp]# ./matrix_native 16 64
./matrix_native: error while loading shared libraries: libiomp5.so: cannot open shared object file: No such file or directory
[root@mic4-mic0 /tmp]# export LD_LIBRARY_PATH=/tmp
[root@mic4-mic0 /tmp]# echo $LD_LIBRARY_PATH
/tmp
[root@mic4-mic0 /tmp]# ./matrix_native 16 64
THREAD_NUM=16
N=64
[root@mic4-mic0 /tmp]#
```

Step 4: Start idbc_mic
Start idbc_mic using the following command:

```
idbc_mic -tco -rconnect=tcpip:coprocessor-ip-address:port
```

For example:

```
[root@mic4 matrixMul]# idbc_mic -tco -rconnect=tcpip:172.31.1.1:2000
idbserver_mic   100% 1609KB    1.6MB/s   00:00
Intel(R) Many Integrated Core Architecture Debugger (Intel(R) MIC Debugger) Version 13.0, Build [78.833.23]
```

Step 5: Configure the searching path (optional)
The user can configure the searching path for the library to load the files, which are needed. The command format for this is:

```
set solib-search-path path[:path]
```

For example:

```
(idb) set solib-search-path /tmp
(idb) set solib-search-path /lib64
(idb) set solib-search-path /sharedir_mic3/wuqing/mic_samples/matrixMul/
```

Step 6: Load debugging program
There are two methods to load debugging:
1. Load executable program directly
Load the executable program through file command. Use the run command to start the program.
2. Debug special process of MIC
Log into the special MIC device through SSH and enter the directory storing executable files.

```
[root@mic4 ~]# SSH mic0
[root@mic4-mic0 /root]# cd /tmp
[root@mic4-mic0 /tmp]# ls
coi_procs       idbserver_mic   libiomp5.so     matrix.c        matrix_native
[root@mic4-mic0 /tmp]#
```

Start the MIC program using the command-line style method:

```
[root@mic4-mic0 /tmp]# ./matrix_native 16 64
THREAD_NUM=16
N=64
```

View the process ID of the native mode running program on MIC using the ps command:

```
[root@mic4-mic0 /tmp]# ps -ef |grep matrix_native
19274 root          0:54 ./matrix_native 16 64
19276 root          0:00 grep matrix_native
[root@mic4-mic0 /tmp]#
```

Use the attach command to track the offload_main process on MIC, in which pid is the process ID which we mentioned in the last step.

```
attach    pid   /opt/intel/composerxe_mic/compiler/lib/mic/offload_main
```

For example:

```
(idb) attach 19274 /sharedir_mic3/wuqing/mic_samples/matrixMul/matrix_native
Attaching to program: /sharedir_mic3/wuqing/mic_samples/matrixMul/matrix_native, process 19274
[New Thread 19274 (LWP 19274)]
Reading symbols from /sharedir_mic3/wuqing/mic_samples/matrixMul/matrix_native...done.
attach_idb () at /sharedir_mic3/wuqing/mic_samples/matrixMul/matrix.c:17
17              } while (loop);
(idb) l
12
13      __declspec(target(mic)) void attach_idb() {
14              volatile int loop = 1;
15              do {
16                      volatile int a = 1;
17              } while (loop);
18      }
19      void matrixMul_cpu(float *A, float *B, float *C, int width)
20      {
21              int i,j,k;
(idb)
```

Step 7: Set breakpoint. See offload mode for details.

Step 8: Modify the loop value to jump out of the infinite cycle. See offload mode for details.

Step 9: Debug the MIC program. See offload mode for details.

6.3   MIC Profiling Tool VTune

Intel VTune Amplifier XE is a high-performance profiling tool for a multi-threading development. It helps one to analyze the sequential and parallel features for C/C++, .NET, and Fortran developers. It supports plenty of performance analysis information, assists the developers in finding bottlenecks, and then in increasing the potential performance of new multi-core processors.

VTune is extremely useful for judging or locating the following issues:

- Finding the most time-consuming functions in the codes
- Finding the code sections that are not fully utilizing computational resources
- Finding the sequential and multi-threading codes with the most potential for optimization
- Finding the synchronous operations weakening the application performance
- Deciding if/where/why the I/O operation takes too much time
- Reporting the variations in performance effects between different synchronous methods, different threading numbers, or different algorithms.
- Processing activities and communications
- Hardware bottlenecks of the codes

On the MIC, the Intel VTune amplifier XE supports analyzing Lightweight Hotspots and collecting some hardware events. Analyzing the user-defined performance model based on hardware events is also supported.

VTune works in two steps:

1. Collecting: collecting runtime hardware events
2. Analyzing: analyzing the runtime characters of the applications based on the collected events, which is also referred to as performance analysis.

Next, we will introduce the methods of using VTune on MIC:

1. Compile codes in the debug mode

   The option –g is used to compile codes in order to make the collected results related to the source codes. It is convenient for viewing and analyzing.

2. Start VTune

   VTune supports two interfaces: GUI (graphics user interface) and CommandLine mode.

   a. VTune GUI: The GUI interface (Fig. 6.30) is started by the command amplxe-gui.

   b. VTune CommandLine: CommandLine interface is started by the command amplxe-cl.

   VTune Command supports writing scripts that are used for nohup/auto collecting.

   To simplify operations, VTune GUI supports using GUI and CommandLine together. Users can configure the collecting parameters for VTune projects using the GUI interface (Fig. 6.31) and then clicking the CommandLine button to view the related CommandLine of the current configuration (Figs. 6.32 and 6.33).

   Since VTune CommandLine can be viewed easily through the GUI, this section only focuses on theVTune GUI. Please read the official manual for VTune CommandLine for further information.

3. Create and Configure a project

   a. Create a VTune project. Configure the project name and location (Fig. 6.34).

   b. Select the target to analyze for the VTune project. Open the VTune project configuration window. Configure the VTune project properties.

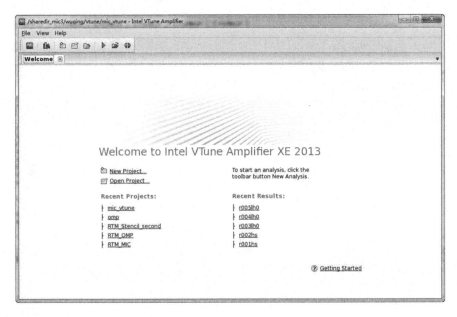

Fig. 6.30   Start VTune GUI

First, select the target to analyze. VTune supports three methods of target analysis: selecting the executable program directly, selecting the process, and system-level performance analysis. In this section, we mainly introduce the first two methods.

 i.   Select the executable program as the analyzing target. In the Application column, we can select the executable program by exact location (Fig. 6.35).

      Fill in the running options within the box containing the Application parameters (Fig. 6.36).

 ii. Select the process as the analyzing target. We can select the process using its name or ID (Fig. 6.37).

      The Start button will be available in the VTune main window with the successfully chosen configuration of the two methods which we mentioned above (Fig. 6.38). Otherwise, if it fails, the Start button turn gray (i.e., if the executable files are not available or the process to be analyzed is finished), and the configuration window shows the error information (Fig. 6.39).

 c. Add searching directories. We can add source code directories, head file directories, and library directories into the VTune project (Figs. 6.40 and 6.41).

 d. Modify project configuration. Users can modify the VTune project configuration if necessary. Click the Project Properties button on the right side of the

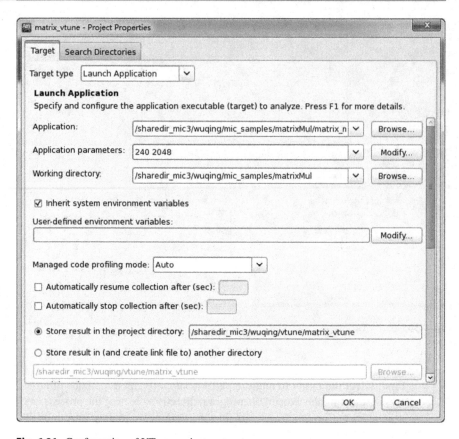

Fig. 6.31 Configuration of VTune project parameters

VTune main window to open the configuration window and configure the project.

   e. Select the analysis type. VTune shows all the analysis types on the left column, including both general and hardware-specialized types. The user-defined analysis type can also be found.

   f. General analysis types: algorithm analysis, including lightweight hotspot analysis, hotspot analysis, parallelism analysis, locking and waiting analysis, etc. (Fig. 6.42).

   g. Hardware-specialized analysis type: VTune needs the processor's hardware support when collecting hardware-architecture-related analysis. This type includes memory access (I/O) bandwidth, average (micro)instruction delaying, and caching rate. Figure 6.43 shows the MIC special analysis types that VTune currently supports.

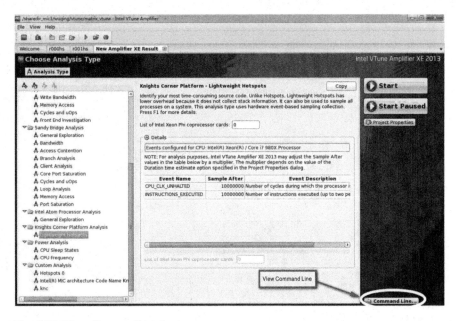

Fig. 6.32 View CommandLine button

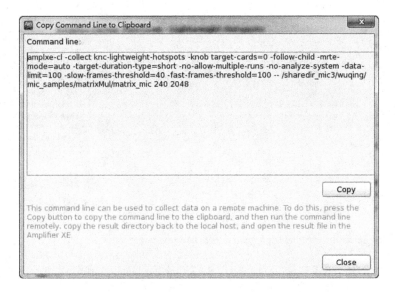

Fig. 6.33 View CommandLine related to the current project

> h. User-defined analysis type: The user can create a customized analysis type based on the VTune supported events (including general events and special hardware events, such as MIC-supported events).
>
> i. Next, we will show how to create a MIC analysis type.

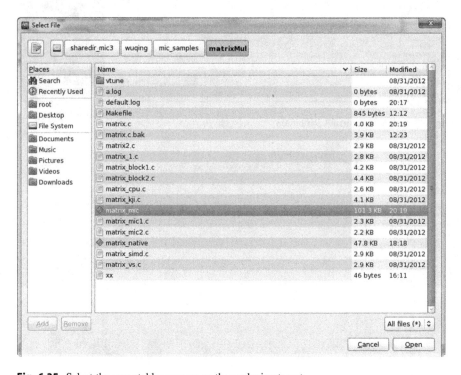

Fig. 6.34 Configure the project name and location

Fig. 6.35 Select the executable program as the analyzing target

    i. Right-click Custom Analysis. Select the hardware platform and event type (Fig. 6.44). The created project is shown as Fig. 6.45.

   ii. Configure user-defined analysis type. Click Edit button for the configuration.

      Add the hardware event to be collected (see the official manual for hardware-supported events). See Fig. 6.46.

      Select ID of the target device (MIC for this example). See Fig. 6.47.

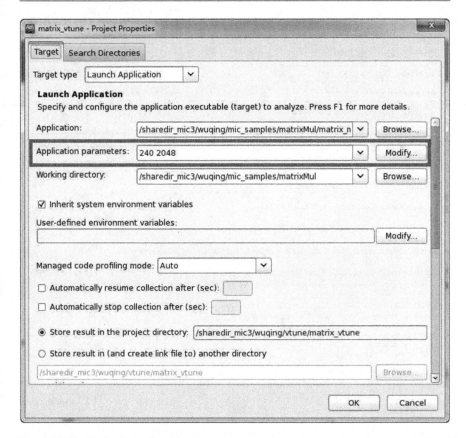

Fig. 6.36 Configure the application parameters

    iii. Start collecting application runtime data. Click the Start button on the right column of the VTune main window to start the collection process. The collecting time can be configured in the project. The collection process can be paused, resumed, and stopped using the Pause, Resume, and Stop buttons, respectively (Fig. 6.48).

    iv. View and analyze the collection results. VTune gives the result interface automatically, as shown in Fig. 6.49.

        VTune supports multi-level profiling crossing code, function, library files, and the full program's running time, with a variety of viewing modes, including percentage mode and exact mode. In addition, VTune also has a sort function and can relate the analysis result to source codes and viewing function trees. Figure 6.50 shows the profiling result in general, including total time consumption, main hotspot function names, and times. Figure 6.51 shows the Top-down Tree, which displays the functions' timing results and the entire timing result. The profiling

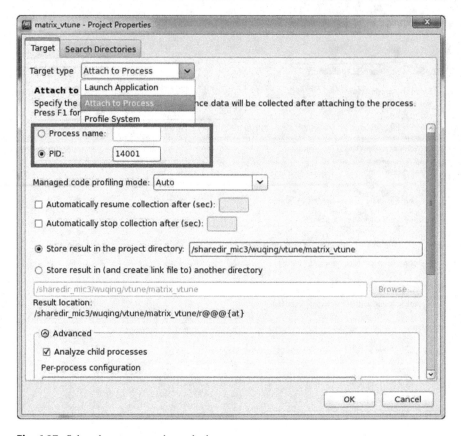

Fig. 6.37 Select the process as the analyzing target

result can be related to specific source code lines. User can click the
function name to view it (Fig. 6.52).

v. According to the profiling result, we can find the performance bottlenecks
and optimize the codes. This is a cycle of "optimizing–profiling–
optimizing".

In this section, we introduced a possible way to profile MIC programs using Intel
VTune Amplifies XE, which is an efficient tool for profiling MIC codes. Users can
download the VTune manual from Intel's official website for more details. We
think more and more profiling tools will begin to support MIC in order to match the
rapid spread in the use of MIC architecture. Besides the profiling tools, the user can
also add time functions manually as in the traditional method for CPU optimizing,
which is a simple way to time exactly each function and code section.

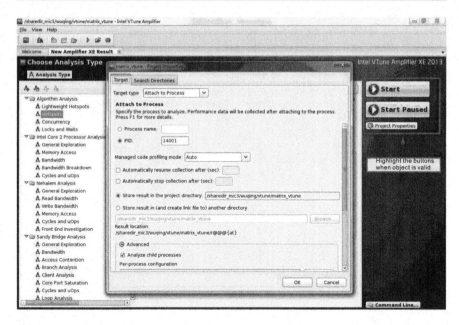

Fig. 6.38 Configuration works

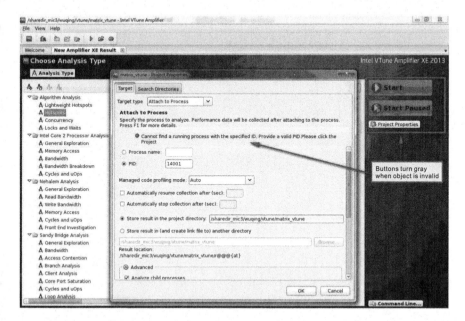

Fig. 6.39 Configuration fails

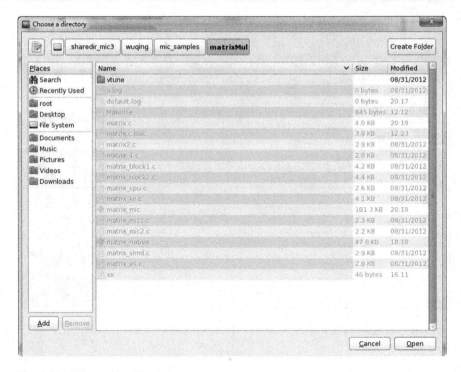

Fig. 6.40 Add searching directories

Fig. 6.41 How to add searching directories

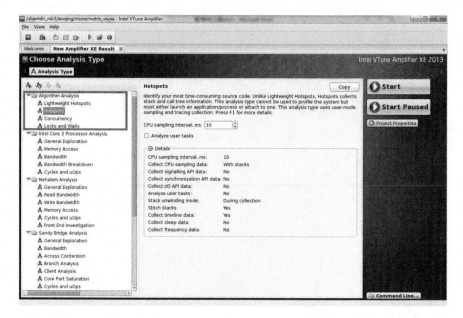

Fig. 6.42 General analysis types

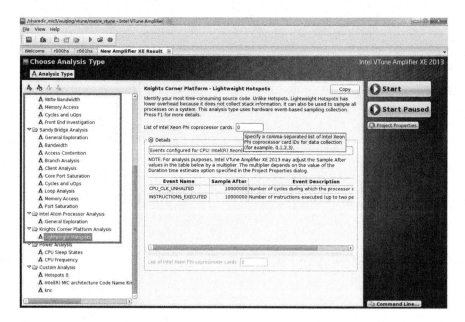

Fig. 6.43 Hardware-specialized analysis type

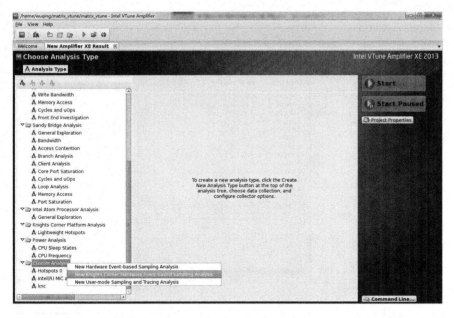

Fig. 6.44 Create user-defined analysis type

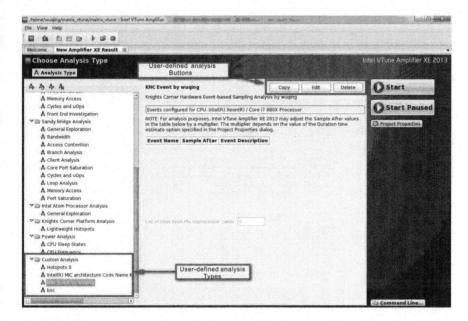

Fig. 6.45 User-defined analysis type

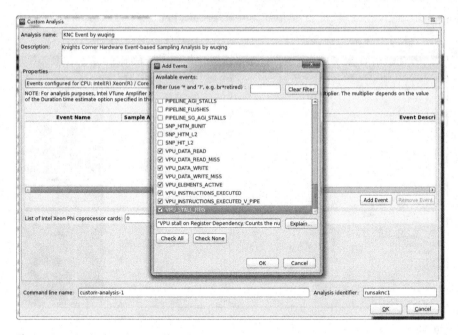

Fig. 6.46 Add hardware events for collecting

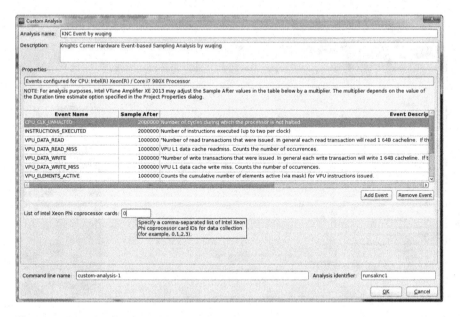

Fig. 6.47 Select ID of target device

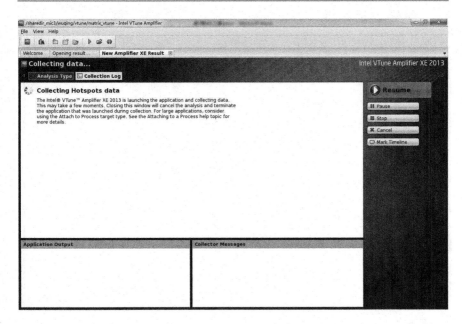

Fig. 6.48 Start of data collecting

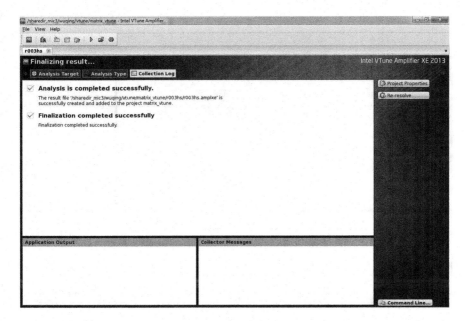

Fig. 6.49 VTune processes the collected results when the collection stops

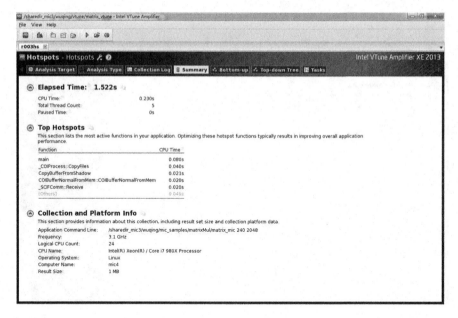

Fig. 6.50 Profiling results in general

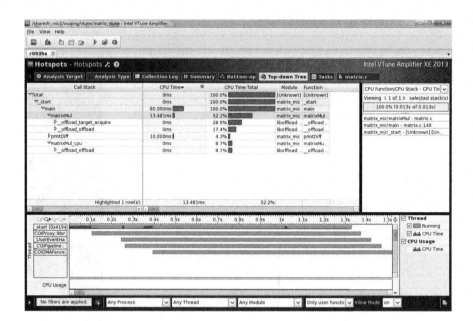

Fig. 6.51 A description of the Top-down Tree

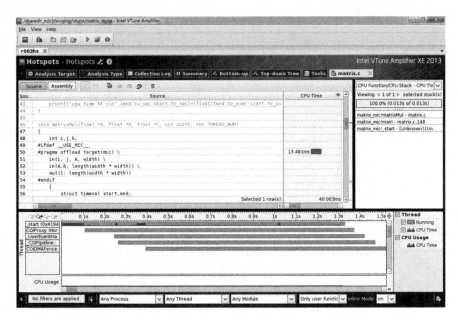

Fig. 6.52 Relating the profiling result to the source codes

Intel Math Kernel Library

7

In order to achieve optimal performance on multi-core and multi-processor systems, we need to fully use the features of parallelism and manage the memory hierarchical characters efficiently. The performance of sequential codes relies on the instruction-level and register-level SIMD parallelism, and also on high-speed cache-blocking functions. Threading applications need advanced planning to achieve satisfactory load balancing.

Normally, using an optimized threading library is an easy way to add parallelism for high-performance computing. It can really reduce the programming time, the necessary tests, and assessments. Moreover, the standard API can guarantee the portability.

Chapter Objectives.

- Review the MKL mathematical library
- Study how to use the Intel Math Kernel Library (MKL) on MIC

7.1   Introduction to the Intel Math Kernel Library

The Intel Math Kernel Library (MKL) is designed to help programmers take advantage of high-performance computing (HPC) using multi-cores, multi-processors, or clusters. Intel MKL has many functions to achieve this on IA-32 or Intel 64 platforms, using Windows, Linux, or Mac OS X systems.

Intel MKL supports a series of optimized and threaded math functions to fully use all the Intel processor's features. When calling functions for the first time, MKL starts the runtime checking to know which hardware is to be utilized. According the results of the check, MKL selects suitable functions to take advantage of instruction-level and register-level SIMD parallelism. Intel MKL has also integrated threadsafe functions to guarantee the functions work normally.

Intel MKL is based on Intel C++ and Fortran by using OpenMP in the threading. This library can fully utilize the multi-cores and multi-processors considering load

E. Wang et al., *High-Performance Computing on the Intel® Xeon Phi™*,
DOI 10.1007/978-3-319-06486-4_7, © Springer International Publishing Switzerland 2014

Math Library	Applications
Basic Linear Algebra Subprograms (BLAS)	All the dense and sparse matrix operations (level-3) are on oriented to be threaded for BLAS. Many vector operations (level-1) and matrix with vector operations are oriented to be threaded on Intel 64 architecture for dense matrices. For sparse matrix, all the level-2 operations, except the triangular sparse matrix solver, are threaded.
Linear Algebra PACKage (LAPACK)	Parts of computing examples are threaded for the following problems: linear algebra equations solver, orthogonal factorization, singular value decomposition, and symmetric eigenvalue problems. Because LAPACK calls BLAS, nonthreaded functions can also be run in parallel.
Scalable LAPACK (ScaLAPACK)	Distributed memory parallel LAPACK oriented to clusters.
PARDISO	All the three steps of this parallel direct sparse matrix solver are threaded: reordering (optional), factorizing, and solving (with multiple right-side projects)
Fast Fourier Transform (FFT)	FFT is used for signal processing, the oil industry, and medical imaging.
Threaded FFT	All the parts of FFT, except with one-dimensional real number and complex number splitting, are threaded.
Cluster FFT	Cluster FFT is applicable to distributed parallel FFT on clusters.
Vector Math Library (VML)	Arithmetic, triangular functions, exponent/logarithm, divisor, etc.

Fig. 7.1 Intel MKL

balance. Figure 7.1 shows the math applications with threading (for Intel MKL 10.2 Update 3):

The MKL functions listed above are all supported on the MIC. However, the following libraries are optimized oriented to the MIC card and are therefore more efficient.

- BLAS (Level 1, 2, and 3), Sparse BLAS, and LAPACK
- VML (Vector Math Library) and VSL (Vector Statistical Library), including Random Number Generators (RNG)
- FFT

Next, we focus on the methods using the MKL on MIC and also on FFT and BLAS.

7.2   Using Intel MKL on MIC

Considering that most HPC users work on the Linux platform, all the examples below are based on Linux. Before starting, we should check to verify whether MKL is installed and the environmental variables are configured.

1. Intel MKL library is installed at <Composer XE directory>. Check to see if the subdirectory of <Composer XE directory> (or namely <mkl directory>) is set up.
2. Check whether the following files exist in <mkr directory>/bin:
   Mklvars.sh
   Mklvars.csh
   intel64/mklvars_intel64.sh
   intel64/mklvars_intel64.csh

Next we configure the environmental variables:
Once the Intel MKL (including the MIC architecture) is installed successfully, select intel64/mklvars_intel64.csh or mklvars.csh (Cshell), intel64/mklvars_intel64.sh or mklvars.sh (Bash and Boume shell) to set the environmental variables. Run shmklvars_intel64.sh or sh mklvars.sh intel64 mod to finish it. This finishes the configuration.

The key point of this chapter is not to introduce how to use Intel MKL functions. Rather, we will focus on how to use the Intel MKL on MIC. Intel MKL supports two libraries to use MIC: one is an Intel 64 host-based library (Linux); the other is a MIC-based library. We can apply two ways to control Intel MKL offload computing functions on MIC: compiler-aided offload mode and automatic offload mode. Next, we discuss these methods.

7.2.1   Compiler-Aided Offload

Data transfer is carried out by using #pragma offload target (mic) in(...) out(...) inout(...){ }. The content in braces calls the MKL library for computation. On MIC, functions are helped immensely by the compiler. The library functions are offloaded to MIC in this manner. We do not need to call MKL functions for MIC explicitly, but just the functions with the same name. The compiler decides the platform on which to run them. Next, we use an example to explain this idea.

The following C code (mkl_test.cpp) is an example of using VML library on MIC. It shows the process of calculating the logarithm of 10.

```
[code]
#include<mkl.h>
#include<mkl_service.h>
#include<stdio.h>
#include<math.h>
#define N (10)
1    int main()
2    {
3        double* datad_a = new double[N];
4        double* datad_b = new double[N];
5    #pragma offload target(mic:0) inout(datad_a,datad_b:length(N))
6    {
7        for(inti=0; i<=N; i++ )
8        {
9            datad_a[i] = i+1;
10           datad_b[i] = log10(i+1);
11       }
         vdLog10(N,datad_a,datad_b); // Call MKL function to calculate log10(datad_a)
12   }
13
14       for(inti=0;i<N;i++)
15   {
16           printf("% 25.14f % 25.14e\n",datad_a[i],datad_b[i]);
17   }.
18   }
```

Some header files should be included. The difficulty is to link the correct library when compiling. Different CPU libraries are used: one is –offload-option; the other is –offload-attribute-target=mic (attribute it as MIC). Please compile this program with the following commands:

```
ipc  mkl_test.cpp  -I/opt/intel/mkl/include  -L/opt/intel/mkl/lib/intel64  -lmkl_intel_ilp64  -lmkl_intel_thread
-lmkl_core     -liomp5     -lm     -offload-attribute-target=mic     -no-intel-extensions     -offload-build
-offload-option,mic,compiler,"-L/opt/intel/mkl/lib/mic/  -lmkl_intel_ilp64  -lmkl_intel_thread  -lmkl_core
-liomp5"
```

The running result is shown as:

```
 1.00000000000000      0.00000000000000e+00
 2.00000000000000      3.01029995663981e-01
 3.00000000000000      4.77121254719662e-01
 4.00000000000000      6.02059991327962e-01
 5.00000000000000      6.98970004336019e-01
 6.00000000000000      7.78151250383644e-01
 7.00000000000000      8.45098040014257e-01
 8.00000000000000      9.03089986991944e-01
 9.00000000000000      9.54242509439325e-01
10.00000000000000      1.00000000000000e+00
```

The result above is exactly same as that on the CPU.

When using the MIC for computing, multi-thread methods such as OpenMP are always employed. We can also use the following code, which is no different from CPU programming.

```
#pragma omp parallel private(…) num_threads(num)
```

7.2.2   Automatic Offload Mode

Automatic offload mode lets MKL functions run on MIC without explicitly writing offload instructions. Under the default situation, Intel MIC MKL can decide how to allocate resources for CPU and MIC. We can also do it manually.

To start offloading automatically to control the computing speed, we can set environmental variables or call functions.

1. Set the environmental variables

First, we must make sure the environmental variable MIC_LD_LIBRARY_PATH includes the path <mkl_directory>/lib/mic, which has a library to automatically offload Intel MKL.

Use a shell script to set up the environmental variables:

a. for Bash shell:

```
export MKL_MIC_ENABLE=1
export MKL_MIC_WORKDIVISION=<value>
export MKL_MIC_WORKDIVISION<number>=<value>
```

b. for C shell:

```
set MKL_MIC_ENABLE=1
set MKL_MIC_WORKDIVISION=<value>
set MKL_MIC_WORKDIVISION<number>=<value>
```

<value> denotes the computing rate on MIC, which varies between 0.0~1.0.
<number> denotes the MIC ID.

2. Call functions

Besides setting environmental variables, we can also start offload automatically on the MIC by the use of calling functions. Table 7.1 shows the related variables and functions.

Table 7.1 Related environmental variables and functions

Environmental variable	Functions	Description
MKL_MIC_ENABLE	mkl_mic_enable	Start offload automatically
MKL_MIC_WORKDIVISION	mkl_mic_set_workdivision	If the offload mode is started automatically, it is used for computation resource division on the MIC
MKL_MIC_WORKDIVISION<number>	mkl_mic_set_workdivision	If the offload mode is started automatically, it is used for computation resource division on MIC and CPU

By calling the functions mentioned above, MIC can make use of the special speed of computation to call MKL. Next, we show an SGEMM example to show how to use this automatic offload mode.

```
[code]
1     /* System headers */
2     #include <stdio.h>
3     #include <stdlib.h>
4     #include <malloc.h>
5     #include <stdint.h>
6     #include "mkl.h"
7
8     int main(int argc, char **argv)
9     {
10        float *A, *B, *C; /* Matrices */
11        MKL_INT N = 2560; /* Matrix dimensions */
12        MKL_INT LD = N; /* Leading dimension */
13        int matrix_bytes; /* Matrix size in bytes */
14        int matrix_elements; /* Matrix size in elements */
15
16        float alpha = 1.0, beta = 1.0; /* Scaling factors */
17        char transa = 'N', transb = 'N'; /* Transposition options */
18
```

```
19       int i, j; /* Counters */
20
21       matrix_elements = N * N;
22       matrix_bytes = sizeof(float) * matrix_elements;
23       A = malloc(matrix_bytes);
24       if (A == NULL) {
25           printf("Could not allocate matrix A\n");
26           return -1;
27       }
28       B = malloc(matrix_bytes);
29       if (B == NULL) {
30           printf("Could not allocate matrix B\n");
31           return -1;
32       }
33       C = malloc(matrix_bytes);
34       if (C == NULL) {
35           printf("Could not allocate matrix C\n");
36           return -1;
37       }
38       for (i = 0; i < matrix_elements; i++) {
39           A[i] = 1.0; B[i] = 2.0; C[i] = 0.0;
40       }
41
42       printf("Computing SGEMM on the host\n");
43       sgemm(&transa, &transb, &N, &N, &N, &alpha, A, &N, B, &N, \
44   &beta, C, &N);
45   printf("Enabling Automatic Offload\n");
46       /* Alternatively, set environment variable MKL_MIC_ENABLE=1 */
47       if (mkl_mic_enable() != 0)
48       {
49           printf("Could not enable Automatic Offload (no MIC devices?). "
50                       "Exiting.\n");
51               return -1;
52       }
53       else
54       {
55               int ndevices = mkl_mic_get_device_count(); /* Number of MIC devices */
56               printf("Automatic Offload enabled: %d MIC devices present\n\n",
57                           mkl_mic_get_device_count());
58               printf("Computing SGEMM with automatic workdivision\n\n");
59               sgemm(&transa, &transb, &N, &N, &N, &alpha, A, &N, B, &N,&beta, C, &N);
60
```

```
61              for (i = 0; i < ndevices; i++)
62              {
63                    /* Alternativelly, set environment variable
64                     * MKL_MIC<i>_WORKDIVISION=1.0 */
65                    printf("Setting workdivision for device MIC:%02d to 1.0\n", i);
66                    mkl_mic_set_workdivision(MKL_TARGET_MIC, i, 1.0);
67                    printf("Resulting workdivision configuration:\n");
68                    double workdivision;
69                    mkl_mic_get_workdivision(MKL_TARGET_HOST,0, &workdivision);
70                    printf("\tworkdivision[HOST] = %+4.2f\n", workdivision);
71                    for (j = 0; j < ndevices; j++)
72                    {
73                          mkl_mic_get_workdivision(MKL_TARGET_MIC,j, &workdivision);
74                          printf("\tworkdivision[MIC:%02d] = %+4.2f\n", j, workdivision);
75                    }
76
77                    printf("Computing SGEMM on device %02d\n\n", i);
78                    sgemm(&transa, &transb, &N, &N, &N, &alpha, A, &N, B, &N, \
79                          &beta, C, &N);
80              }
81        }
82        free(A); free(B); free(C);
83        printf("Done\n");
84        return 0;
85  }
```

The above complete code shows how to call functions to automatically start offload. In the main function, the array space is allocated and initialized first (see lines 10–40 for detail). To compare the CPU codes, we do the sgemm computation on the CPU first (lines 43–44). From line 47, the code determines the offload automatically. For this to work, it uses mkl_mic_set_workdivision() to assign tasks for MIC, in which the parameter MKL_TARGET_MIC/HOST is used to assign the task object, then to do the sgemm computation. In this step, mkl_mic_get_workdicision() is used to obtain the division rate.

The compiling commands are the following:

```
icc -c -O3 -openmp -I/opt/intel/composer_xe_2013.0.079/mkl/include sgemm.c -o sgemm.o
iccsgemm.o  -openmp  -L/opt/intel/composer_xe_2013.0.079/mkl/lib/intel64 -lmkl_intel_lp64 -lmkl_intel_thread
-lmkl_core -lpthread-lm -o sgemm.out
```

The running result is:

```
Computing SGEMM on the host
Enabling Automatic Offload
Automatic Offload enabled: 1 MIC devices present
Computing SGEMM with automatic workdivision
Setting workdivision for device MIC:00 to 1.0
Resulting workdivision configuration:
        workdivision[HOST] = -1.00
        workdivision[MIC:00] = +1.00
Computing SGEMM on device 00
Done
```

Generally speaking, the compiler-aided offload mode is needed more for modifying codes than the automatic offload mode. However, compiler-aided offload can better control the data transfer, which also helps to improve the optimization.

7.3   Using FFT on the MIC

7.3.1   Introduction to FFT

Fast Fourier Transform (FFT) is widely used in the fields of acoustics, telecommunications, electricity, image and signal processing, exploration geophysics, radar, satellites, medical imaging, etc. FFT implementation belongs to the area of digital technology. FFTW is a free toolkit released by MIT, written with the C programming language, to calculate one-dimensional or multi-dimensional discrete Fourier transforms. The FFTW generator is written with Caml, an object-oriented language. It can be migrated to many hardware platforms. It supports shared memory multi-threading parallelism and distributed MPI parallelism. FFTW is one of the most widely used FFT toolkits. FFTW applies the arbitrary-size transform, which can perform many transform operations. The inner structure is transparent to the users. FFTW's speed is fast (it is suitable for compilers on different machines, and the code generator utilizes AST in the runtime and can be optimized by itself, without taking compiling time.). Many scientific and industrial applications, including quantum physics, spectroscopy, audio signal processing, oil exploration, earthquake prediction, weather forecast, etc., employ FFTW to cope with massive FFT operations.

7.3.2   A Method to Use FFT on the MIC

The Intel MKL has an interface that is compatible with FFTW. Without modifying codes, users can use the Intel MKL by changing some options.

The use of the MIC version of the FFT library is similar to the other functions in MKL, and just needs linking to the different MIC libraries. Below we show how to use the Intel MKL FFT in the compiler-aided offload mode.

1. Source code:

```
[code]
1    #include <stdio.h>
2    #include <math.h>
3    #include <stdlib.h>
4    #include <float.h>
5    #include "fftw3.h"
6
7    __attribute__ ((target(mic)))static void init(fftwf_complex *x, int N, int H, fftwf_complex *x_host);
8    __attribute__ ((target(mic)))static int verify(fftwf_complex *x, int N, int H);
9    __attribute__ ((target(mic)))static void output(fftwf_complex *x, int N, int H, fftwf_complex *x_host);
10
11   int main(void)
12   {
13       int N = 64;
14       int H = -N/2;
15       fftwf_complex *x_host  = fftwf_malloc(sizeof(fftwf_complex)*N);
16       printf("Allocate array for input data\n");
17       //init x_host......
18
19       #pragma offload target(mic:0) inout(x_host:length(N))
20       {
21           /* FFTW plan handles */
22           fftwf_plan   forward_plan = 0, backward_plan = 0;
23           fftwf_complex   *x = 0;
24           x  = fftwf_malloc(sizeof(fftwf_complex)*N);
25           int status = 0;
26
27           printf("Create FFTW plan for 1D double-precision forward transform\n");
28           forward_plan = fftwf_plan_dft(1, &N, x, x, FFTW_FORWARD, FFTW_ESTIMATE);
29           if (0 == forward_plan) goto failed;
30
31           printf("Create FFTW plan for 1D double-precision backward transform\n");
32           backward_plan = fftwf_plan_dft(1, &N, x, x, FFTW_BACKWARD, FFTW_ESTIMATE);
33           if (0 == backward_plan) goto failed;
34
35           printf("Initialize input for forward transform\n");
36           init(x, N, H);
37
38           printf("Compute forward FFT\n");
39           fftwf_execute(forward_plan);
40
41           printf("Verify the result of forward FFT\n");
42           status = verify(x, N, H);
43           if (0 != status) goto failed;
44
45           printf("Initialize input for backward transform\n");
46           init(x, N, -H);
47
48           printf("Compute backward transform using new-array function\n");
49           fftwf_execute_dft(backward_plan, x, x);
50
```

```
51          printf("Verify the result of backward FFT\n");
52          status = verify(x, N, H);
53          output(fftwf_complex *x, int N, int H, fftwf_complex *x_host);
54
55          if (0 != status) goto failed;
56          cleanup:
57
58          printf("Destroy FFTW plans\n");
59          fftwf_destroy_plan(forward_plan);
60          fftwf_destroy_plan(backward_plan);
61
62          printf("Free data array\n");
63          fftwf_free(x);
64          printf("TEST %s\n",0==status ? "PASSED" : "FAILED");
65          goto end;
66   failed:
67          printf(" ERROR\n");
68          status = 1;
69          goto cleanup;
70   end:
71      }
72      //......
73   }
```

The three functions above denote initializing the inputted arrays, verifying the results, and outputting the results, respectively. We need to add __attribute__ ((target(mic))) for the function declaration. Because this part has no relationship with this usage, there is no need to show the source code associated with these subfunctions.

We use fftwf_complex *x_host=fftwf_malloc(sizeof(fftwf_complex)*N) to allocate memory space and initialize the arrays. The arrays are uploaded onto MIC for computation. The result is returned with x_host.

2. The makefile:

```
icc fft_mic_lxw.c -I/opt/intel/mkl/include/fftw -L/opt/intel/mkl/lib/intel64 -lmkl_intel_ilp64 -lmkl_intel_thread
-lmkl_core -liomp5 -lm -offload-attribute-target=mic -no-intel-extensions
-offload-option,mic,compiler,"-L/opt/intel/mkl/lib/mic/ -lmkl_intel_ilp64 -lmkl_intel_thread -lmkl_core
-liomp5"
```

3. The result:

```
Example sp_plan_dft_1d
Forward and backward 1D complex inplace transform
Configuration parameters:
 N = 64
 H = -32
Allocate array for input data
Create FFTW plan for 1D double-precision forward transform
Create FFTW plan for 1D double-precision backward transform
Initialize input for forward transform
Compute forward FFT
Verify the result of forward FFT
Initialize input for backward transform
Compute backward transform using new-array function
Verify the result of backward FFT
Destroy FFTW plans
Free data array
......
TEST PASSED
```

7.3.3   Another Method to Use FFT on the MIC

Intel MKL supports another method to call FFT functions using its own interface. We show an example using Intel's FFT interface rather than the FFTW for the compiler-aided offload mode.

1. An FFT example of using offload on the MIC: complex_dft_1d.c:

```
[code]
1    /* define precision: float or double */
2    #if !defined(REAL)
3    #define REAL float
4    #endif
5
6    /* package the standard pointer using pragma offload_attribute */
7    #if __INTEL_OFFLOAD
8    #include <offload.h>
9    #else
10   #define _Offload_get_device_number() (-1)
11   #endif
12
13   #include <stdio.h>
14   #include <stdlib.h>
15   #include <math.h>
16   #include <float.h>
17   #include <complex.h>
18   #include "mkl_dfti.h"
19
```

```
20   /* Define the macros TARGET_MIC and PRAGMA_OFFLAD */
21   #if __INTEL_OFFLOAD
22   #define TARGET_MIC          __declspec(target(mic))
23   #define PRAGMA_OFFLOAD(args) __pragma(offload args)
24   #else
25   #define TARGET_MIC          /*none*/
26   #define PRAGMA_OFFLOAD(args) /*none*/
27   #endif
28
29   /* total number of MIC cards */
30   TARGET_MIC int micno = 0;
31
32   /* define the COMPLEX type */
33   typedef REAL _Complex COMPLEX;
34
35   /* declare two aid functions */
36   void init(COMPLEX *x, int N, int H);
37   int  verify(COMPLEX *x, int N, int H);
38   /* offload FFT onto card */
39   int demo_fft(int N)
40   {
41       /* running status: nonzero means it fails */
42       MKL_LONG status = 0;
43
44       /* I/O array pointer */
45       static TARGET_MIC COMPLEX *x = 0;
46
47       /* DFTI descriptor */
48       static TARGET_MIC DFTI_DESCRIPTOR_HANDLE hand = 0;
49
50       /* clear before the demo_fft returns */
51       int for_cleanup = 0;
52       const int MIC_HAND = 1;
53       const int MIC_X = 2;
54       const int HOST_X = 4;
55       printf("\nPrepare DFTI descriptor for N=%i on the card\n", N);
56
57       PRAGMA_OFFLOAD(target(mic:micno) nocopy(hand) out(status) in(N))
58       {
59           status = DftiCreateDescriptor(&hand,
60               sizeof(REAL)==sizeof(float) ? DFTI_SINGLE : DFTI_DOUBLE,
61               DFTI_COMPLEX, 1, (MKL_LONG)N );
62           if (0 == status)
63           {
64               status = DftiCommitDescriptor(hand);
65           }
66       }
```

```
67          for_cleanup |= MIC_HAND;
68      if (0 != status)
69      {
70          printf("Error: cannot create descriptor, status=%li (%s)\n",\
71                  (long)status, DftiErrorMessage(status));
72          goto cleanup;
73      }
74      printf("Allocate space for data on the host\n");
75      x = (COMPLEX*)malloc( N * sizeof(COMPLEX) );
76      if (0 == x)
77      {
78          printf("Error: no host memory for %li bytes\n", (long)(N*sizeof(COMPLEX)));
79          status = -1;
80          goto cleanup;
81      }
82      for_cleanup |= HOST_X;
83      printf("Preallocate buffers on the card\n");
84      PRAGMA_OFFLOAD(target(mic:micno) inout(x:length(N) align(4096) alloc_if(1) free_if(0)))
85      {
86          // This might fail with COI_RESOURCE_EXHAUSTED error for large N,
87          // which cannot be intercepted with current offload usage model
88      }
89      for_cleanup |= MIC_X;
90
91      // use the descriptor for computing
92      for (int i = 0; i < 3; ++i)
93      {
94          /* Pick an "arbitrary" harmonic to verify FFT */
95          int H = ((i+1)*N) % 11;
96
97          /* Initialize input for forward transform */
98          init(x, N, H);
99
100         //----------------------------------------------------------------
101         printf("Offload computation onto the card, H=%i\n",H);
102         //----------------------------------------------------------------
103         PRAGMA_OFFLOAD(target(mic:micno) out(status) nocopy(hand)
104             /* Reuse preallocated buffers by specifying alloc_if(0) free_if(0) */
105             inout(x:length(N) alloc_if(0) free_if(0))
106             in(N,H))
107         {
108             /* Compute just forward FFT */
109             status = DftiComputeForward(hand, x);
110             /* Other computation may be done here */
111         }
112         if (0 != status)
113         {
114             printf("Error: computing FFT failed, status=%li (%s)\n",
115                     (long)status, DftiErrorMessage(status));
116             goto cleanup;
```

```
117              /* verify the result */
118              status = verify(x, N, H);
119              if (0 != status)
120              {
121                  printf("Error: incorrect result for N=%i, H=%i\n", N, H);
122                  goto cleanup;
123              }
124          }
125
126      cleanup:
127          printf("Cleanup resources\n");
128          if (for_cleanup & MIC_HAND)
129          {
130              PRAGMA_OFFLOAD(target(mic:micno) nocopy(hand))
131              {
132                  DftiFreeDescriptor(&hand);
133              }
134          }
135          if (for_cleanup & MIC_X)
136          {
137              PRAGMA_OFFLOAD(target(mic:micno) nocopy(x:length(N) alloc_if(0) free_if(1)))
138              {
139              }
140          }
141          if (for_cleanup & HOST_X)
142          {
143              free(x);
144          }
145
146          printf("Test for N=%i %s\n", N, 0==status ? "passed" : "FAILED");
147          return status;
148      }
149
150      char *default_argv[] = { "", "1024", "524288" };
151      int main(int argc, char *argv[])
152      {
153          PRAGMA_OFFLOAD(target(mic:micno) out(micno))
154          {
155              micno = _Offload_get_device_number();
156              char version[DFTI_VERSION_LENGTH];
157              DftiGetValue(0, DFTI_VERSION, version);
158              if (micno < 0)
159                  printf("HOST using %s\n", version);
160              else
161                  printf("MIC%i using %s\n", micno, version);
162              fflush(0);
163          }
```

```
164     if (argc < 2)
165     {
166         argv = default_argv;
167         argc = sizeof(default_argv)/sizeof(default_argv[0]);
168     }
169 /* run demo for each given FFT size */
170     int failed = 0;
171     for (int a = 1; a < argc; ++a)
172     {
173         int N = atoi(argv[a]);
174         failed += demo_fft(N);
175     }
176     printf("\n%s\n", failed ? "Some tests FAILED" : "All tests passed");
177     return failed;
178 }
179 /* exactly calculate (K*L)%M */
180 double moda(int K, int L, int M)
181 {
182     return (double)(((long long)K * L) % M);
183 }
184
185 /* initialize arrays */
186 void init(COMPLEX *x, int N, int H)
187 {
188     double TWOPI = 6.2831853071795864769, phase;
189
190     for (int n = 0; n < N; n++)
191     {
192         phase  = moda(n,H,N) / N;
193         ((REAL*)&x[n])[0] = cos( TWOPI * phase ) / N;
194         ((REAL*)&x[n])[1] = sin( TWOPI * phase ) / N;
195     }
196 }
197 int verify(COMPLEX *x, int N, int H)
198 {
199     double err, errthr, maxerr;
200     int n;
201     const double EPSILON = sizeof(REAL)==sizeof(float)
202         ? FLT_EPSILON : DBL_EPSILON;
203     errthr = 5.0 * log( (double)N ) / log(2.0) * EPSILON;
204     printf(" Verifying the result, errthr = %.3lg\n", errthr);
205
```

```
206    maxerr = 0;
207    for (n = 0; n < N; n++)
208    {
209        double re_exp = 0.0, im_exp = 0.0, re_got, im_got;
210
211        if ((n-H)%N==0)
212        {
213            re_exp = 1;
214        }
215
216        re_got = ((REAL*)&x[n])[0];
217        im_got = ((REAL*)&x[n])[1];
218        err  = fabs(re_got - re_exp) + fabs(im_got - im_exp);
219        if (err > maxerr) maxerr = err;
220        if (!(err <= errthr))
221        {
222            printf(" x[%i]: ",n);
223            printf(" expected (%.17g,%.17g), ",re_exp,im_exp);
224            printf(" got (%.17g,%.17g), ",re_got,im_got);
225            printf(" err %.3lg\n", err);
226            return 100;
227        }
228    }
229    printf(" Verified, maximum error was %.3lg\n", maxerr);
230    return 0;
231 }
```

The kernel part is in the demo_fft function of this example. The main job is to transfer data (line 57) to the MIC device through PRAGMA_OFFLOAD(target (mic:micno) nocopy(hand) out(status) in(N)){ }. DftiCreateDescritptor() is used to create the descriptor. DftiCommitDescriptor(hand) is used to commit the descriptor. DftiComputeForward(hand,x) is used for the forward transform (line 109). Finally, DftiFreeDescriptor(&hand) is used to release the descriptor resource.

2. Compiling options:

```
icc -c -offload-attribute-target=mic -O3 -openmp -std=c99 -DMKL_ILP64
-I/opt/intel/composer_xe_2013.0.079/mkl/include complex_dft_1d.c -o complex_dft_1d.o
 icc complex_dft_1d.o -openmp -Wl,--start-group
/opt/intel/composer_xe_2013.0.079/mkl/lib/intel64/libmkl_intel_ilp64.a
/opt/intel/composer_xe_2013.0.079/mkl/lib/intel64/libmkl_intel_thread.a
/opt/intel/composer_xe_2013.0.079/mkl/lib/intel64/libmkl_core.a -Wl,--end-group -lpthread -lm
-offload-option,mic,compiler,"-Wl,--start-group
/opt/intel/composer_xe_2013.0.079/mkl/lib/mic/libmkl_intel_ilp64.a
/opt/intel/composer_xe_2013.0.079/mkl/lib/mic/libmkl_intel_thread.a
/opt/intel/composer_xe_2013.0.079/mkl/lib/mic/libmkl_core.a -Wl,--end-group" -o complex_dft_1d.out
```

3. Execute /complex_dft_1d.out
    The result follows:

```
MIC0 using Intel(R) Math Kernel Library Version 11.0.0 Product Build 20120801 for Intel(R) 64 architecture
applications

Prepare DFTI descriptor for N=1024 on the card
Allocate space for data on the host
Preallocate buffers on the card
Offload computation onto the card, H=1
  Verifying the result, errthr = 5.96e-06
  Verified, maximum error was 1.91e-08
Offload computation onto the card, H=2
  Verifying the result, errthr = 5.96e-06
  Verified, maximum error was 4.11e-08
Offload computation onto the card, H=3
  Verifying the result, errthr = 5.96e-06
  Verified, maximum error was 2.28e-08
Cleanup resources
Test for N=1024 passed


Prepare DFTI descriptor for N=524288 on the card
Allocate space for data on the host
Preallocate buffers on the card
Offload computation onto the card, H=6
  Verifying the result, errthr = 1.13e-05
  Verified, maximum error was 1.26e-08
Offload computation onto the card, H=1
  Verifying the result, errthr = 1.13e-05
  Verified, maximum error was 1.14e-08
Offload computation onto the card, H=7
  Verifying the result, errthr = 1.13e-05
  Verified, maximum error was 1.7e-08
Cleanup resources
Test for N=524288 passed
All tests passed
```

7.4   Use BLAS on the MIC

7.4.1   A Brief Introduction to BLAS

Basic Linear Algebra Subprograms (BLAS), an advanced toolkit of NetLib, implements the basic linear algebra operations. There are three levels in BLAS: Level-1 is "vector–vector" operations; Level-2 is "matrix–vector" operations; and Level-3 is "matrix–matrix" operation. BLAS is a Fortran-style library for which some Fortran style habits must still be obeyed when it is called by C, such as transferring variables by address, not by value and column–major array.

CBLAS is a C-style interface to BLAS. CBLAS follows the common C-style habits. When using CBLAS, the head file mkl.h defines enum type and the other prototypes of all the functions. The head file simplifies the programmer's workload and decides whether or not to compile the codes by using C++. When one uses C++, the library should also be correct.

7.4.2   How to Call BLAS on the MIC

The only difference between using the BLAS library on the MIC and on the CPU is the linking library. We also give an example showing how to use BLAS on the MIC, by using the compiler-aided offload mode.

```
[code]
/* System headers */
1    #include <stdio.h>
2    #include <stdlib.h>
3    #include <malloc.h>
4    #include <stdint.h>
5
6    /* MKL header */
7
8    #include "mkl.h"
9
10   int main(int argc, char **argv)
11   {
12          float *A, *B, *C; /* Matrices */
13
14          MKL_INT N=5, NP; /* Matrix dimensions */
15          int matrix_bytes; /* Matrix size in bytes */
16          int matrix_elements; /* Matrix size in elements */
17
18          float alpha = 1.0, beta = 1.0; /* Scaling factors */
19          char transa = 'N', transb = 'N'; /* Transposition options */
20
21          int i, j; /* Counters */
22
```

```
23        /* Check command line arguments */
24        if (argc< 2) {
25                printf("\nUsage: %s <N>\n\n", argv[0]);
26        } else {
27        /* Parse command line arguments */
28                N = atoi(argv[1]);
29        }
30
31        if (N <= 0) {
32                printf("Invalid matrix size\n");
33                return -1;
34        }
35
36        printf("\nMatrix dimension is being set to %d \n\n", (int)N);
37
38        matrix_elements = N * N;
39        matrix_bytes = sizeof(float) * matrix_elements;
40
41        /* Allocate the matrices */
42        A = (float*)malloc(matrix_bytes);
43        if (A == NULL) {
44                printf("Could not allocate matrix A\n");
45                return -1;
46        }
47
48        B = (float *)malloc(matrix_bytes);
49        if (B == NULL) {
50                printf("Could not allocate matrix B\n");
51                return -1;
52        }
53
54        C = (float*)malloc(matrix_bytes);
55        if (C == NULL) {
56                printf("Could not allocate matrix C\n");
57                return -1;
58        }
59
60        /* Initialize the matrices */
61        for (i = 0; i <matrix_elements; i++) {
62                A[i] = 1.0; B[i] = 2.0; C[i] = 0.0;
63        }
64
```

```
65   #pragma offload target(mic) \
66         in(transa, transb, N, alpha, beta) \
67         in(A:length(matrix_elements)) \
68         in(B:length(matrix_elements)) \
69         in(C:length(matrix_elements)) \
70         out(C:length(matrix_elements) alloc_if(0))
71         {
72               sgemm(&transa, &transb, &N, &N, &N, &alpha, A, &N, B, &N,\
73               &beta, C, &N);
74         }
75
76         /* Display the result */
77         printf("Resulting matrix C:\n");
78         if (N>10) {
79               printf("NOTE: C is too large, so print only its upper-left 10x10 block...\n");
80               NP=10;
81         } else {
82               NP=N;
83         }
84         printf("\n");
85         for (i = 0; i < NP; i++) {
86               for (j = 0; j < NP; j++)
87                     printf("%7.3f ", C[i + j * N]);
88               printf("\n");
89         }
90
91         /* Free the matrix memory */
92         free(A); free(B); free(C);
93
94         return 0;
95   }
```

The makefile:
1. The compiling option on the CPU can be transferred automatically to the MIC compiler. Using -offload-option, mic, the compiler can pass the MIC options.
2. The library information for compiling MIC libraries should be –offload-option, mic,ld.

```
    icc -c -offload-attribute-target=mic -O3 -openmp -std=c99 -I/opt/intel/composer_xe_2013.0.079/mkl/include
sgemm.c -o sgemm.o
    icc sgemm.o -openmp -L/opt/intel/composer_xe_2013.0.079/mkl/lib/intel64 -lmkl_intel_lp64
-lmkl_intel_thread -lmkl_core -lpthread -lm
-offload-option,mic,compiler,"-L/opt/intel/composer_xe_2013.0.079/mkl/lib/mic -lmkl_intel_lp64
-lmkl_intel_thread -lmkl_core" -o sgemm.out
```

Run the example ./segemm.out 10
The result is as follows:

```
Matrix dimension is   set to 10
Resulting matrix C:
  20.000  20.000  20.000  20.000  20.000  20.000  20.000  20.000  20.000  20.000
  20.000  20.000  20.000  20.000  20.000  20.000  20.000  20.000  20.000  20.000
  20.000  20.000  20.000  20.000  20.000  20.000  20.000  20.000  20.000  20.000
  20.000  20.000  20.000  20.000  20.000  20.000  20.000  20.000  20.000  20.000
  20.000  20.000  20.000  20.000  20.000  20.000  20.000  20.000  20.000  20.000
  20.000  20.000  20.000  20.000  20.000  20.000  20.000  20.000  20.000  20.000
  20.000  20.000  20.000  20.000  20.000  20.000  20.000  20.000  20.000  20.000
  20.000  20.000  20.000  20.000  20.000  20.000  20.000  20.000  20.000  20.000
  20.000  20.000  20.000  20.000  20.000  20.000  20.000  20.000  20.000  20.000
  20.000  20.000  20.000  20.000  20.000  20.000  20.000  20.000  20.000  20.000
```

In order to achieve the best performance of the MKL on the MIC, we should reduce transferring data and increase the computing workload. We should also try to reuse the data on the coprocessor's memory and use special methods, such as nocopy and free_if(0), as much as possible to reduce the memory reallocation time on the MIC and data transfer time.

Part II

Performance Optimization

The optimization methods for MIC programs are introduced and explained in this section through an actual example. After this section, readers will be able to optimize an existing MIC program so that it will make full use of the hardware.

Performance Optimization on MIC

8

This chapter covers MIC performance optimization methods, including optimizing parallelism, memory management, data transfer, accessing memory, vectorization, load balance, and scalability. Many examples are used to demonstrate each optimization method.

Chapter Objectives.
- Understand the performance optimization strategies on the MIC.
- Understand general methods of optimizing MIC codes through improving parallelism, memory management, vectorization, and caching performance.
- Understand how to use nocopy, asynchronous, and SCIF technologies to optimize communication between the CPU and the MIC.
- Understand how to optimize for load balance and thread scalability on the MIC.

8.1 MIC Performance Optimization Strategy

MIC performance optimization includes two parts: the first is optimizing the communication between the CPU and the MIC, and the second is optimizing the computing kernel and the memory access. Optimization on the MIC is a cyclic process, as presented in Fig. 8.1, and includes the following steps:

1. Obtain profiling data as the basic reference
2. Analyze the profiling data and find the bottleneck through VTune
3. Decide the optimization strategy according to the bottleneck.
4. Optimize the codes.
5. Test the result. If the result is correct and the performance is improved, an optimization cycle is finished.
6. Decide if the next optimization cycle should begin. The factors to be considered include: Has the performance been improved? Has the MIC hardware resource been fully used?

E. Wang et al., *High-Performance Computing on the Intel® Xeon Phi™*,
DOI 10.1007/978-3-319-06486-4_8, © Springer International Publishing Switzerland 2014

Fig. 8.1 Performance
optimization cycle

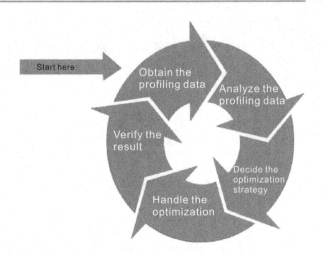

The common keypoints we should focus concerning MIC performance optimization are listed below.

1. Select appropriate test cases, i.e., the workload. The appropriate workload needs to satisfy some requirements such as testability, repeatability, and stability. The testing time should not be too long or too short, the tests should be repeatable, the results should be stable without large error, and the test case should not be too different from the common ones. For example, if the selected test case uses path 1 of the algorithm while the other test runs go through path 2, then the selected case is not appropriate.

2. Obtain performance indicators. The easiest and simplest indicator is time consumption. We can use the time function to calculate the kernel time consumption and the data-transfer time consumption. Or use VTune to obtain each thread's time consumption on the MIC cores.

3. Consider the hotspots and key paths. We should focus on optimizing the key points, such as the basic line, the limitation, and the potential of the computing performance. When facing these key points that limit MIC performance, indicators such as GFLOPS, CPI, parallelism, data locality, bandwidth, vectorization, and IO bottlenecks should be seriously considered. These indicators can be measured through VTune.

4. In the process of optimization, we also need to consider the coding quality, such as the portability, readability, maintainability, and reliability. We can modify the compiling options and math libraries, and manually modify the hotspot codes, etc.

5. The test case should cover all the situations we might meet. The correct result is all we need to test for MIC performance indicators.

6. Finally, we should decide if we should continue on to the next cycle of optimization.

The MIC program's performance optimization includes two levels: system level and kernel level. System-level optimization includes the optimization of load balance between different nodes, as well as between the CPU and MIC. It also includes memory optimization, parallelism optimization between computation and IO as well as IO and IO, data transfer optimization, network optimization, hard disk optimization, and so on. Kernel-level optimization covers the optimization of parallelism, load balance, the synchronization of processes and threads, thread scalability, vectorization, cache, data alignment, and the selection of library functions. Next, we introduce these optimization methods in detail.

8.2 MIC Optimization Methods

As a many-core coprocessor, MIC can only achieve high performance if all of these cores are fully being utilized. The objective of parallelism is to make the program run on each core at the same time in order to improve the performance. Two types of parallel programs are data parallelism and task parallelism.

Data parallelism is based on data partition. It has good scalability: with a significant increase of data scale and threads, the workload of each thread does not decrease. Thus, data parallelism is the natural style. Note that uneven partitioning will cause a serious load unbalance. MIC is a shared, memory-based many-core coprocessor, so data parallelism is a must.

Task parallelism is based on multi-task parallel operation. There are many different methods to deal with different tasks, and there is no dependency on data. Thus, for the co-computing between CPU and MIC, task parallelism is feasible. Both the CPU and MIC can deal with different tasks to fully play each of their individualized roles.

When running parallel programs, the application will become very complex for existing multi-task/multi-thread communications. To achieve the best performance on MIC, all aspects should be optimized.

8.2.1 Parallelism Optimization

In computer architecture, parallelism means the maximum number of parallel running instructions. We can simplify parallelism as the number of threads or processes that can simultaneously run on the multi- or many-core processor. For the same program, a different parallelism design will cause a tremendous impact on performance. On the MIC, the parallelism is related to the number of threads or processes, parallel levels, parallel granularity, etc.

8.2.1.1 Parallelism

The MIC card has many physical cores, and each core can launch four threads, so the programmer needs to design as many threads as possible to fully use all these cores. For example, on a 60-core MIC, we can launch 240 threads, so the best number of threads for each core is 3 or 4. Figure 8.2 shows the performance

Fig. 8.2 An example of performance scalability on MIC

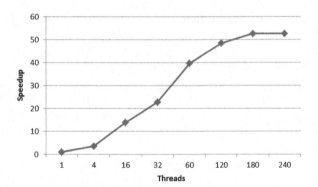

scalability on a 60-core MIC card by setting different threads, from which we can see that only by fully using all of the cores can we improve the MIC performance perfectly. However, the number of threads should be appropriate, because with too many threads the resource consumption is also tremendous. Therefore, set the threads to guarantee parallelism and high utilization of the MIC.

8.2.1.2 Parallel Granularity

Whether the parallel program selects the appropriate level is the key point to be considered for optimization. According to the principle of coarse-grained parallelism, we should optimize the codes at the top level. Parallelization on the outer level can benefit programming and also the performance: an increase of granularity decreases thread management and thread resources. It is especially important for the MIC performance when launching hundreds of threads.

Next, we give a simple example to show parallel levels. There are two levels in the codes and no dependency on data, so both levels can be parallelized. We apply the two-level cycle parallel method to implement coarse-grained parallelism.

```
     [pseudo-code]
1.   #pragma omp parallel for num_threads(THREAD_NUM)
2.   for (i=0; i<M; i++)
3.   {
4.       for (j=0; j<N; j++)
5.       {
6.           ...
7.       }
8.   }
```

We note here that not all of the applications are appropriate for outer-level cycle parallelism, which may cause the thread access to span too far, leading to a cache miss. In this situation, inner level cycle parallelism is a better choice. At the same time, we can launch multiple threads before the outer level "for", and do the task threading in the inner level "for".

```
[pseudo-code]
1   #pragma omp parallel num_threads(THREAD_NUM)
2   for (i=0; i<M; i++)
3   {
4   #pragma omp for
5       for (j=0; j<N; j++)
6       {
7           ...
8       }
9   }
```

In practice, the application may not satisfy the MIC parallelism in some cycles. In this situation, we can apply the multi-level cycle merging method. For example, in the codes above, $M = 20$ and $N = 30$, so any level of parallelization cannot meet the MIC parallelism requirement. We can merge the two levels of 'for' so that the cycle number becomes 600, which then satisfies the MIC's requirement. The merging method is shown below:

```
[pseudo-code]
1   #pragma omp parallel for num_threads(THREAD_NUM)
2   for (k=0; k<M*N; k++)
3   {
4       i = k/M;
5       j = k%M;
6       ...
7   }
```

Alternatively, nested parallelism is also feasible:

```
[pseudo-code]
1   omp_set_nested(true); //Allow nested parallelism
2   #pragma omp parallel for num_threads(THREAD_NUM1)
3   for (i=0; i<M; i++)
4   {
5   #pragma omp parallel for num_threads(THREAD_NUM2)
6       for (j=0; j<N; j++)
7       {
8           ...
9       }
10  }
```

8.2.2 Memory Management Optimization

As a coprocessor, MIC memory space is very limited compared to the host, and it isn't expandable. Thus, fully using the limited memory space while supporting more computation has become MIC optimization's main problem. In addition, because of the card memory space limitation, some programs also encounter a great challenge in MIC portability.

High parallelism may also bring new problems to memory space. Temporary memory space for each iteration step in previous sequential programs is reusable. After developing parallel CPU programs, temporary space needs to be allocated independently for each thread. But the number of computing units is few, and there are also concurrent threads. Memory problems are not prominent in this situation. However, concurrent threads can reach 200 or more on the MIC, leading to a sharp increase in memory space requirements. Assuming that one thread (private variable) takes 1 MB of memory, the total memory consumption is 4 MB or 8 MB on the CPU, while it can expand to more than 200 MB according to the number of threads on MIC. So optimizing memory becomes an inevitable challenge.

In addition, each MIC core's clock frequency is lower than that of the CPU. Therefore, there are some differences between the devices and the CPU regarding the efficiency of building memory space. So MIC memory space optimization not only works in the space but also in time.

As mentioned above, MIC memory optimization includes two parts: memory usage capacity and allocation times. Memory usage capacity means the amount of memory the program uses, usually for the maximum usage during the run. Allocation times are the number of times the memory is opened statically or dynamically during the runtime of the program. When optimization focuses on the usage capacity, sometimes it will generate conflict with other optimization methods focusing on performance. But the optimization of memory occupancy is generally connected with the portability, so we usually sacrifice some of the performance in exchange for optimizing space.

8.2.2.1 Memory Occupancy

When memory occupancy becomes a bottleneck, there are generally two scenarios. One is the task itself needing a huge amount of memory, and there are difficulties in porting due to the small memory size on the MIC. The other is that each thread takes a lot of temporary space when porting onto the MIC, so having many threads leads to insufficient memory. The methods for reducing memory occupancy include:

1. Task Partition

The basic method of solving memory occupancy is task partition. If we divide a big task into small tasks and decrease the memory requirements at the same time, we can solve any memory insufficiency problems. For different situations, we can take different task partition methods, which are now discussed.

In the first case, when a task itself occupies too much space, generally the task itself needs to be partitioned, processing one subset of the task at a time.

In the second case, when each thread needs a lot of temporary space, we generally choose to reduce the degree of concurrency and the number of threads; then the total memory occupancy can be cut down. Because of the many threads on MIC, reducing the concurrency to some extent can still keep performance in an acceptable range.

If the task can't be divided, or if memory in the MIC can't meet the smallest partition requirement, we have to consider that such a program (generally considered as an example) is not appropriate to be ported to MIC. Even on the host, the memory is limited. If the task is too big, it cannot even run on the CPU. Strictly speaking, in such a situation, the program is not portable and just needs to be divided into smaller parts. Although each part is only a fragment, the constantly repeating process of transmission–calculation–transmission–calculation, etc., can finish the whole task. However, in most cases, performance brought by such a method can't make up the rapid expansion in development costs. This method serves only as a reference for those extreme cases.

2. Temporary space reuse

Some temporary space used by some program can be merged and saved. For instance, the size of array a is 100 MB, used in the first half of the program, and the size of array b is 150 MB, used in the latter part of the program. When b is in use, the variable a is no longer used. So it only needs to allocate 150 MB of space for both a and b. Although decreasing the temporary space can bring adverse effects for the readability of codes, if it has a strong impact on performance, the balance between maintainability and performance should be taken care of, and an equilibrium point can be found.

There is another case, shown below:

```
[code snippet]
1    for(i=0;i<N;++i)
2    {
3        c[i]=a[i]+N;
4        d[i]=b[i]+N;
5    }
```

The variables a and b are intermediate variables, while c and d are arrays to get. The variables a' c and b' d don't have a dependence relationship, so the loop can be completely divided into two: one for calculating c, the other for calculating d. At the end of calculating c and the beginning of calculating d, the space of array a can be released to save memory occupancy.

3. Changing the algorithm

In the view of program algorithms, coarse-grained parallelism (also known as higher parallel-level circulation) generally consumes more memory resources, while fine-grained parallelism takes less. For example, a calculation of N matrix multiplication $A*B$ application has a large memory if we calculate the N matrix multiplication in parallel. But memory occupancy can be decreased if we parallelize the calculation of each $A*B$ matrix, and sequentially calculate the N matrices. The difference between this method and the task partition is that it should change the algorithm (the function of one thread was formerly a serial calculation of matrix multiplication, and now it is a calculation of one element of the matrix) and the granularity of parallelism, and task partition generally only needs to change the size of the task and transfer method.

In the view of details of coding algorithms, there is a classical interview question: how to change the value of two integer variables without using a temporary variable? Although this method may not be suitable on the MIC, it points us in the direction of trying to find another algorithm to save memory and achieve the same functionality.

In the traditional methods from the "ancient times" of programming history, the hardware resources were deficient, so programmers tried their best to save even 1 bit of space. With increased development in hardware, however, those skills are gradually forgotten or never used. Nowadays, the coprocessor presents the same problem. Therefore, some skills and methods may be found in "ancient" books and references, which we leave to the reader to review.

8.2.2.2 Allocation Times

The optimization of memory allocation times does not necessarily reduce memory occupancy, but the performance can be improved with some of these techniques. For performance optimization in terms of memory space, the most important point is to put the operation that is opening up space outside of the loop. Whether we are talking about a simple loop or a subfunction called by the loop, both can open private space themselves based on their needs, and an especially large memory space can be opened by using the function malloc. Because of MIC's clock frequency, the operation of allocating space is slower than that of the CPU. Therefore, if we allocate larger memory inside the loop, each operation will be delayed. Although each allocation takes just a little more time, together they can cause overall performance loss in some cases. For instance, we see this problem in the process of some software porting optimization. The code fragment is very simple: a loop which calls computing functions to allocate memory space first and then do computation. Our task is to port the loop onto the MIC.

The porting itself is simple, but we find that the computing function only takes half of the entire running time. The first suspect is the offload transfer, but the problem doesn't reappear when we use our written test case. In the latter test, we discover that the problem is due to allocating space inside the computing function. Because the time for each allocation is considerable, and the allocation is according to the loop, the running time rapidly increases. If we move the memory allocations outside the loop and just do one memory allocation together, the program's running time can be significantly shortened. In the process of transferring memory allocation, by declaring a pointer in the host and using nocopy to allocate space in offload, we can further save some unnecessary transfer time. To determine the size of the allocated space, we take the number of threads multiplied by the size of the memory that a single cycle needs, and each thread searches its private memory address according to its own thread ID.

When we need to call the offload function many times to perform a series of operations, if there is a common array in different offload functions, we can instead apply it only once by using the method of nocopy to use it multiple times. This reduces data transfer time, but it also avoids applying space many times. This type of data transfer optimization method is introduced in detail in the next section.

8.2.3 Data Transfer Optimization

Data transfer is a process of communication, which greatly affects the parallel computing performance. For single-node computing, frequent send and receive operations bring damage to parallel performance. For cluster computing, the huge communication consumption has a fatal impact. The communication between processors takes most of the time during the execution of parallel programs, so we need to reduce I/O as much as possible.

In general, the data transfer optimization methods include nocopy, offload asynchronous, SCIF mode, and 4K multiple. Now, we introduce these methods.

8.2.3.1 The nocopy Method
The CPU and MIC communicate through PCI-E, which is fairly slow. Thus, we need to reduce the communication workload by using nocopy technology.

In the offload mode, the statements "in" and "out" allocate and release the memory at the beginning and end of each offload by default. The data and memory, however, can be reused in many applications, so the reallocation operations are not necessary all of the time. The nocopy technology can be utilized in applications that call offload many times, especially those with iterative calls.

An example of MIC code without nocopy is shown as follows:

```
1     [pseudo-code]
2     ...
3     p_c = ...; // the value p_c doesn't change in each iteration
4     for(i=0; i<steps; i++)// many iterations
5     {
6           p_in = ...; // the value p_in changes in each iteration
7
8     #pragma offload target(mic) \
9           in(p_in:length(...)) \
10          in(p_c: length(...)) \
11          out(p_out: length(...))
12        {
13               kernel(p_in, p_c, p_out);
14        }
15    }
16    ... = p_out; // p_out is used only when the iterations on CPU are finished
```

In the code above, p_in, p_c, and _out allocate memory on the MIC, transfer data between the CPU and the MIC, and release memory at each iteration (Fig. 8.3). In practice, p_c only has to be passed to MIC once, and p_out only has to be returned to the CPU after the iterations finish. Apparently, there are a lot of communications waste between the CPU and the MIC. In this situation, we can use nocopy to reduce the communication workload.

Normally, nocopy is jointly used with alloc_if(), free_if(). We have already introduced the usage of nocopy in Chap. 5. Here, we give the MIC code by using nocopy:

```
      [pseudo-code]
1     ...
2     p_c = ...; // the value p_c doesn't change in each iteration
3     #pragma offload target(mic) \
4     in(p_c: length(...) alloc_if(1) free_if(0)) \
5     nocopy(p_in:length(...)alloc_if(1) free_if(0)) \
6     nocopy(p_out: length(...) alloc_if(1) free_if(0))
7     {
8     } //allocate memory   without release;  pass the value of p_c
9     for(i=0; i<steps; i++)// many iterations
10    {
11         p_in = ...; // the value p_in changes in each iteration
12    #pragma offload target(mic) \
13          in(p_in:length(...) alloc_if(1) free_if(0) ) \
14        nocopy(p_c) \
15        nocopy(p_out)    // pass the vaule of p_in in each iteration, using nocopy for p_c and p_out
16        {
17               kernel(p_in, p_c, p_out);
18        }
19    }
20    #pragma offload target(mic) \
21    nocopy (p_c: length(...) alloc_if(0) free_if(1)) \
22    nocopy(p_in:length(...)alloc_if(0) free_if(1)) \
23    out(p_out: length(...) alloc_if(0) free_if(1))
24    {
25    } //return the value of p_out to CPU, and release the allocated memory on MIC
26    ... = p_out; // p_out is used only when the iterations on CPU are finished
```

Fig. 8.3 Flow chart of MIC
program execution

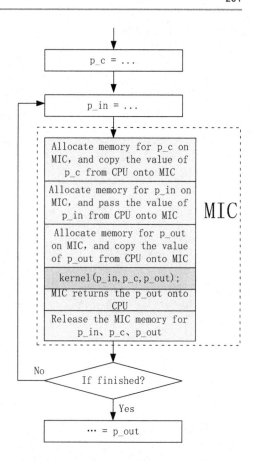

The flow chart for program execution using nocopy is shown in Fig. 8.4, from which we can see that using nocopy technology can reduce the communication times between the CPU and the MIC. For applications with iterations, nocopy can improve the performance significantly.

8.2.3.2 The Offload Mode for Asynchronous Transfer

For the offload mode on MIC, asynchronous transfer means: 1. asynchronous data transfer between the MIC and the host, and computation on MIC; 2. asynchronous computation on the MIC and the host.

1. Asynchronous data transfer and computing

Asynchronous data transfer is frequently used when calling MIC functions many times and no dependency exists between the neighbor functions. This situation happens because the original algorithm, for some reasons such as not having enough

Fig. 8.4 Flow chart of MIC
program with nocopy

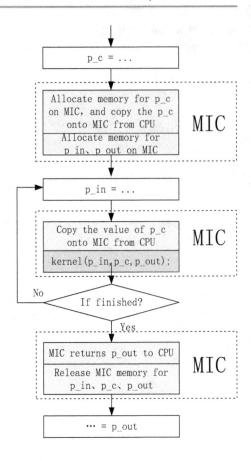

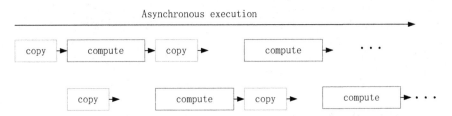

Fig. 8.5 Asynchronous execution diagram

memory on MIC, partitions the data so that the MIC functions without dependency
need to be called many times. This asynchronous method can hide most of the data
transfer time by transferring and computing through an "assembly line" (Fig. 8.5).

Thus, we can optimize the codes using asynchronous data transfer when dealing
with an algorithm by partition, or when data transfer takes so much time that it
affects the execution efficiency.

The data transfer optimization involves two offload statements: offload_transfer and offload_wait. The offload_transfer is used to transfer data and send a signal when finished; its parameters are almost the same as traditional offload statements. The only difference is that no codes follow the offload_transfer. The offload_wait is used to pause the execution until the signal from offload_transfer is received. An example is the following:

```
[code snippet]
//C/C++
1     #pragma offload_attribute(push, target(mic))
2     int count = 25000000;
3     int iter = 10;
4     float *in1, *out1;
5     float *in2, *out2;
6     #pragma offload_attribute(pop)
7
8     void do_async_in()
9     {
10            int i;
11    #pragma offload_transfer target(mic:0) \
12            in(in1 : length(count) alloc_if(0) free_if(0) ) signal(in1)
13
14            for (i=0; i<iter; i++)
15            {
16                    if (i%2 == 0) {
17
18    #pragma offload_transfer target(mic:0) if(i!=iter-1) \
19                            in(in2 : length(count) alloc_if(0) free_if(0) ) signal(in2)
20
21    #pragma offload target(mic:0) nocopy(in1) \
22                            wait(in1) out(out1 : length(count) alloc_if(0) free_if(0) )
23
24                            compute(in1, out1);
25
26                    } else {
27
28    #pragma offload_transfer target(mic:0) if(i!=iter-1) \
29                            in(in1 : length(count) alloc_if(0) free_if(0) ) signal(in1)
30
31    #pragma offload target(mic:0) nocopy(in2) \
32                            wait(in2) out(out2 : length(count) alloc_if(0) free_if(0) )
33
34                            compute(in2, out2);
35                    }
36            }
37    }
```

```fortran
[code snippet]
!Fortran
1     integer, parameter :: iter = 10
2     integer,parameter :: count = 25000
3     !dir$ options /offload_attribute_target=mic
4     real(4), allocatable :: in1(:), in2(:), out1(:), out2(:)
5     integer :: sin1, sin2
6     !dec$ end options
7
8     contains
9
10    subroutine do_async_in()
11       integer i
12
13    !dir$ offload_transfer target(mic:0) &
14       in(in1 :  alloc_if(.false.) free_if(.false.) ) signal(sin1)
15
16       do i = 0, (iter - 1)
17          if (mod(i,2) == 0) then
18
19    !dir$ offload_transfer target(mic:0) if(i/=iter-1) &
20          in(in2 :   alloc_if(.false.) free_if(.false.) ) signal(sin2)
21
22    !dir$ offload target(mic:0) nocopy(in1) wait(sin1) &
23          out(out1 : length(count) alloc_if(.false.) free_if(.false.) )
24
25    call compute(in1, out1);
26
27       else
28
29    !dir$ offload_transfer target(mic:0) if(i/=iter-1) &
30          in(in1 : alloc_if(.false.) free_if(.false.) ) signal(sin1)
31
32    !dir$ offload target(mic:0) nocopy(in2) wait(sin2) &
33          out(out2 : length(count) alloc_if(.false.) free_if(.false.) )
34
35    call compute(in2, out2);
36
37       endif
38    enddo
39
40    end subroutine do_async_in
```

This example only shows the function do_async_in, which is enough to show the application of asynchronous transfer. We have allocated two array spaces in this example. While waiting to use in1 and out1 for computing in the first space, we perform the data transfer in the second space for in2 and out2. After finishing data transfer in the second space and computing in the first space, the second space starts computing and the first space starts transferring data. Run this cycle until all the work is finished. Because the two tasks execute alternately, the total time consumption equals the longer one of the asynchronous processes, and not the sum of the two.

Before starting the cycle, we use offload_transfer to fill the first space (C: lines 11–12, Fortran: lines 13–14). Thus, when the cycle starts (C: line 14, Fortran: line 16), filling the second space can begin right away (C: lines 18–19 Fortran: 19–20). Wait until the transfer in the first space finishes (C: lines 21–22, Fortran: 22–23). Do the computing in the first space afterwards (C: line 24, Fortran: line 25). Because there is only one MIC card, the computing is executed serially. Only the execution that happens between transfer and computation is asynchronous. We should note that, because the example is not a complete code, we haven't allocated space in the first transfer. We already allocated space on MIC outside of this function. In this way, we avoid reallocating space in the cycle and save significant time.

Another point we should note here is the signal parameter. In C/C++, the parameter should be passed with a pointer; in Fortran, it's an integer variable. When handling different programming languages, the signal can only use one parameter, which means in C/C++, no matter how many arrays are transferred at the same time, using only one as the signal parameter is enough.

In optimizing asynchronous data transfer, the main problem is the memory occupancy of the card. For spaces of data transferring and computing simultaneously, those with too large a data partition may not have enough memory space to use. Thus, even for long time transfers, it may not be appropriate to use too much array space for asynchronous processes.

2. Asynchronous computing

Asynchronous computing is often used for CPU/MIC collaborative computing. The CPU and MIC share the responsibility to fully use the computing resources of the node. Normally, when we mention collaborative computing, the CPU and MIC use different threads to perform parallel computing. But this section denotes another way in which they are in the same thread: start the functions in serial and execute in parallel.

In the traditional offload mode, after calling offload/MIC functions, the MIC is in charge while the CPU enters the awaiting state. In the asynchronous computing mode, after calling offload functions, the driver starts the MIC and returns the control back to the CPU thread so that the work on the CPU is still going on. When the MIC functions are finished, it sends a signal to the CPU thread. In the asynchronous mode, the CPU thread and MIC functions are executed in parallel as part of the runtime, which reduces time consumption.

Compared with traditional collaborative computing, this is a fairly simple method without starting multiple threads. It has a better flexibility, even with the slight loss of parallelism. The computing functions on the MIC do not need big changes. The CPU functions can do some data preparation or other work while waiting for the next loop. Traditional collaborative computing cannot achieve this.

Besides the traditional offload statements, asynchronous computing may also use offload_wait to wait for the signal. We give a simple example below:

```
[code snippet]
1     int counter;
2     float *in1;
3     counter = 10000;
4     __attributes__((target(mic))) mic_compute;
5     while(counter>0)
6     {
7     #pragma offload target(mic:0) signal(in1)
8         {
9             mic_compute();
10        }
11        cpu_compute() // parallel execute this function and MIC functions above
12    #pragma offload_wait target(mic:0) wait(in)
13        counter--;
14    }
```

```
[code snippet]
1     integer signal_var
2     integer counter
3     counter = 10000
4     !DIR$ ATTRIBUTES OFFLOAD:MIC :: mic_compute
5     do while (counter .gt. 0)
6         !DIR$ OFFLOAD TARGET(MIC:0) SIGNAL(signal_var)
7             call mic_compute()
8         call cpu_compute() ! parallel execute this function and MIC functions above
9             !DIR$ OFFLOAD_WAIT TARGET(MIC:0) WAIT (signal_var)
10        counter = counter - 1
11    end do
12    end
```

We can see that after calling mic_compute, the CPU controls the program immediately. The CPU can continue executing cpu_compute and wait at offload_wait until MIC finishes its computing work.

8.2.3.3 SCIF Transfer Optimization

In the previous chapters, we have already discussed that the communication between the CPU and the MIC is through PCI-E and PCI-E is fairly slow. Thus, we need to reduce the communication between the CPU and the MIC. But sometimes, such communications cannot be avoided. In these situations, we can consider using SCIF, which can improve the efficiency of frequent transfers of small-size data. An example is outlined below:

Codes on the host (as a requester):

```
[code]
1    #include <stdio.h>
2    #include <stdlib.h>
3    #include <stdint.h>
4    #include <unistd.h>
5    #include <fcntl.h>
6    #include <string.h>
7    #include <sys/ioctl.h>
8    #include <scif.h>
9    #include <sys/time.h>
10   int main(int argc, char **argv)
11   {
12       scif_epd_t epd;
13       int err = 0;
14       int req_pn = 11;
15       int con_pn;
16       struct scif_portID portID;
17       int i, num_loops = 0, total_loop = 10;
18       char *senddata;
19       char *recvdata;
20       int msg_size;
21       int block;
22       int node;
23       struct timeval tv;
24       struct timeval tv1;
25       if (argc != 4) {
26           printf("Usage ./scif_connect_send_recv* <msg_size><0/1 for noblock/block><node>\n");
27           exit(1);
28       }
29       msg_size = atoi(argv[1]);
30       block = atoi(argv[2]);
31       node = atoi(argv[3]);
32       portID.node = node;
33       portID.port = 10;
34       printf("Open the scif driver\n");
35       if ((epd = scif_open()) < 0) {// start an endpoint at the request side
36           printf("scif_open failed with error %d\n", (int)epd);
37       exit(1);
38       }
39       printf("scif_bind to port 11\n");
40   if ((con_pn = scif_bind(epd, req_pn)) < 0) {// requester side: binding this endpoint to its own port
41       printf("scif_bind failed with error %d\n", con_pn);
42       exit(2);
43       }
44       printf("req_pn=%d",req_pn);
45       retry:
```

```
46      if ((scif_connect(epd, &portID)) != 0) {//use retry for communiction with listener. If it's prepared at
the listener side, connection works; otherwise retry will be repeated. If with connection, then go on with the data
transfer operations.
47          if (ECONNREFUSED == errno) {
48              printf("scif_connect failed with error %d retrying\n", errno);
49              goto retry;
50          }
51          printf("scif_connect failed with error %d\n", errno);
52          exit(3);
53      }
54      printf("scif_connect success\n");
55      printf("node=%d,port=%d\n",portID.node,portID.port);
56      while (num_loops < total_loop) {
57          senddata = (char *)malloc(msg_size);
58          if (!senddata) {
59              perror("malloc failed");
60              err = ENOMEM;
61          }
62          memset(senddata, 0x25, msg_size);
63          err = 0;
64          gettimeofday(&tv,0);
65          while ((err = scif_send(epd, senddata, msg_size, block))<= 0) {// here are connections between
the two endpoints, the requester side will send data to listener
66              if (err < 0) {
67                  printf("scif_send failed with err %d\n", errno);
68                  fflush(stdout);
69                  free(senddata);
70                  goto close;
71              }
72          }
73          gettimeofday(&tv1,0);
74          printf("err=%d",err);
75          printf(" total = %f\n", (tv1.tv_sec-tv.tv_sec)*1e6+(tv1.tv_usec-tv.tv_usec));
76          err = 0;
77          free(senddata);
78          num_loops++;
79      }
80  close:
81      printf("Close the scif driver\n");
82      close(epd);
83      if (!err) {
84          printf("Test success\n");
85      }
86      else
87          printf("Test failed\n");
88      return (err);
89  }
```

Codes on MIC (as a listener):

```
[code]
1    #include <stdio.h>
2    #include <stdlib.h>
3    #include <stdint.h>
4    #include <unistd.h>
5    #include <fcntl.h>
6    #include <string.h>
7    #include <sys/ioctl.h>
8    #include <scif.h>
9    int main(int argc, char **argv)
10   {
11       int epd;
12       int newepd;
13       int err = 0;
14       int req_pn = 10;
15       int con_pn;
16       int backlog = 16;
17       struct scif_portID portID;
18       int i, num_loops = 0, total_loop = 10;
19       char *senddata;
20       char *recvdata;
21       int msg_size;
22       int block;
23       if (argc != 3) {
24           printf("Usage ./scif_accept_send_recv* <msg_size><0/1 for noblock/block>\n");
25           exit(1);
26       }
27       msg_size = atoi(argv[1]);
28       block = atoi(argv[2]);
29       portID.node = 2;
30       portID.port = 11;
31       if ((epd = scif_open())< 0) {//try to start endpoint at the listener side
32           printf("scif_open failed with error %d\n", errno);
33           exit(1);
34       }
35       printf("scif_bind to port 10\n");
36       if ((con_pn = scif_bind(epd, req_pn))< 0) {//Listener: binding the builtup endpoint to its own port
37           printf("scif_bind failed with error %d\n", errno);
38           exit(2);
39       }
40       printf("scif_listen with backlog of 16\n");
41       if ((scif_listen(epd, backlog))< 0) {//Listener prepares for listening, and sets the max connection
     number for listener.
42           printf("scif_listen failed with error %d\n", errno);
43           exit(3);
44       }
```

```
45        printf("scif_accept in syncronous mode\n");
46        if ((((scif_accept(epd, &portID, &newepd, SCIF_ACCEPT_SYNC))< 0) && (errno != EAGAIN))
          {//listener accepts the connection request from requester, and sets the sychronous/asychronous mode of
          acceptance
47            printf("scif_accept failed with errno %d\n", errno);
48            exit(4);
49        }
50        printf("scif_accept complete\n");
51        printf("node=%d,port=%d",portID.node,portID.port);
52        while (num_loops < total_loop) {
53            recvdata = (char *)malloc(msg_size);
54            if (!recvdata) {
55                free(senddata);
56                perror("malloc failed");
57                err = ENOMEM;
58            }
59            memset(recvdata, 0x00, msg_size);
60            err = 0;
61            while ((err = scif_recv(newepd, recvdata, msg_size, block))<= 0) {//accept the data from
          requester
62                if (err < 0) {
63                    printf("scif_recv failed with err %d\n", errno);
64                    fflush(stdout);
65                    free(senddata);
66                    free(recvdata);
67                    goto close;
68                }
69            }
70            printf("err=%d",err);
71            err = 0;
72            free(recvdata);
73            num_loops++;
74        }
75    close:
76        printf("Connection is complete\n");
77        fflush(stdout);
78        scif_close(newepd);
79        fflush(stdout);
80        if (!err) {
81            printf("Test success\n");
82        }
83        else
84            printf("Test failed\n");
85        scif_close(epd);
86        return (err);
87    }
```

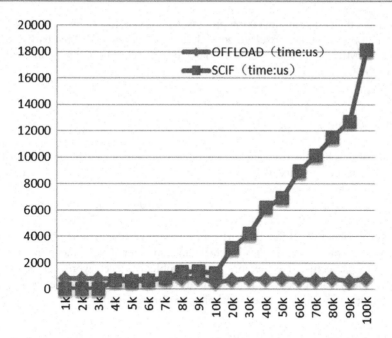

Fig. 8.6 Performance comparison between SCIF and offload

In this example, the CPU runs as the requester and the MIC runs as the listener. When the listener obtains the send request from the related port, it establishes the connection with the requester. This method is also appropriate for different MIC cards. In Chap. 5, where we introduce SCIF, we learned that the SCIF is very good at transferring small-scale data. Based on this example, we did some experiments, the results of which are shown in Fig. 8.6.

When the data size is smaller than 4K, the SCIF performance is better than that of offload. In particular, when the data size is 1K, 2K, and 3K, the performance of the SCIF is 80 times better than that of offload. When the data size is 4K–6K, the performance of the SCIF is almost the same as that of the offload mode. When the data size is bigger than 7K, the SCIF performance declines; when the data is bigger than 10K, there is a significant difference between SCIF and offload. From these results, readers can know when to use SCIF to improve performance. We can also conclude that offload mode runs more stable than SCIF mode.

NOTE: Run the MIC program first to build a listener at the endpoint. When the host program runs, it sends a connection request to this endpoint. It is inconvenient to run two programs on both the server and customer sides. Next, we take the communication between the CPU and the MIC as an example and try to explain how to write these two codes together, and only run them on the customer side. Below are the basic ideas for a single program to use SCIF mode, which benefits from the coding format:

1. Package the two communication processes as unique functions.
2. Launch two separate threads to control the CPU part and the MIC part.
3. Run the MIC part first in order to build the listener to receive a connection request.
4. Run the CPU part to accept the request.

8.2.4   Memory Access Optimization

Memory access has always been the bottleneck of computer architecture. The slow development of memory does not match the rapid development of modern processors, which leaves the processor waiting for memory access latency; this is called the memory wall problem. During the past 10 years, processor speeds have increased by 50%–100% annually while memory speeds have increased by only 7% per year. We can predict that this difference will only become more obvious, so the memory system is still the key obstacle in computer development. In parallel computer architecture, the memory access problem is even worse, especially in the MIC many-core processor with hundreds of threads. Therefore we need to pay more attention to memory access optimization for MIC.

8.2.4.1 MIC Hierarchical Memory

The MIC card is not only similar to the CPU in core architecture – based on x86 – but also in the memory hierarchical structure. MIC applies two levels of caches: the KNC (Knights Corner) chip framework in Fig. 8.7 and the MIC hierarchical structure in Fig. 8.8.

The KNC card has eight dual-channel GDDR5 memory controllers with a bandwidth of 5.5 GT/s, and two levels of cache: L1 and L2. Each core of KNC has a 32KB L1 instruction cache and a 32KB L1 data cache. The L1 cache line is 64B, using 8 channels and 8 banks. The L1 cache belongs to each core privately, with fast accessing speed. KNC has shared L2 caches. The L2 cache on each core has an L1 data cache and an instruction cache. A large, shared L2 cache (31 MB) is organized by these cores. Because each core has a 512KB L2 cache and 62 cores, the L2 cache reaches 31 MB, so it looks like the 31 MB L2 cache is available for memory space. However, if both or multiple cores share data, these shared data on the L2 caches of different cores are duplicated. Without shared data from the codes, the total size of the L2 on chip is 31 MB. Conversely, if each core shares the data or codes simultaneously, the total size of the L2 is 512 KB (each L2 cache stores the same 512 KB data).

8.2.4.2 Memory Access Optimization Strategy on MIC

On MIC, there are mainly two ways to optimize the memory access:
1. Hiding memory latency:

   The basic law for hiding memory latency is to overlap the computing operations when latency occurs in memory access. Two ways of hiding memory latency on MIC are given below:

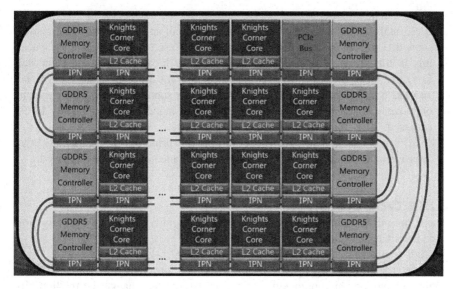

Fig. 8.7 Structure of KNC chip

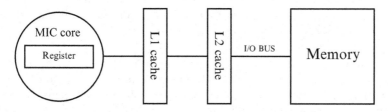

Fig. 8.8 Hierarchical memory on MIC

a. Multi-threading: multi-threading is based on exposing thread-level parallelism to hide memory latency. A classical case is simultaneous multi-threading technology, which uses the basic idea that a suitable thread can be launched in case the current thread hangs onto memory access. Each core on a MIC card supports four threads, using hardware multi-threading technology to hide memory latency. So for MIC programming, launching three to four threads per core can guarantee good performance.

b. Prefetching: with prefetching technology, the data and instructions are fetched before they are needed. Current prefetching technology can be divided into three types: hardware prefetching (extending memory management system architecture), software prefetching (using nonblock, prefetching instructions of modern processors), and mixed prefetching. MIC supports hardware prefetching, which is controlled by MIC hardware automatically.

2. Using Cache

Normally, memory access only involves a small space. Experimental results show that 90% of memory access operations only use 10% of the memory space; this is the principle of locality. Two common locality types are Temporal Locality and Spatial Locality. Temporal Locality means some space is accessed and has a good chance of being accessed again soon. Spatial Locality means once some space is accessed, its neighborhood has a great opportunity to be accessed.

MIC includes L1 and L2 caches. From the principle of locality and the improvement of cache hit rate, memory performance can also be improved.

8.2.4.3 Cache Optimization

Cache optimization mainly uses the principle of locality. There are two ways for code level Cache optimization.

1. Code transform

Code transform means a transformation of program instructions. Most compilers support this optimization. Code transform can change the relationship of the instructions, optimize the instruction locality, improve the instruction Cache performance, and also optimize data locality by changing the execution order of the instructions. Code transform utilizes loop transform in the following methods:

a. Loop fusion

Loop fusion is a code transform method fusing multiple loops. In the process of fusion, the same data can be reused in one iteration of a loop instead of multiple uses in multiple loops. In this way, the temporal locality as well as the data cache performance are improved. Loop fusion expands the loop size, which is good for dispatching instructions. In loop fusion, the original multiple loops may depend on each other. Thus, the difficulty of loop fusion is obeying this dependency.

An example of loop fusion:

```
[code snippet]//original loop:          [code snippet]//Fused loop
    for(i=0;i<n;i++)                        for(i=0;i<n;i++)
        a[i]=b[i]+1;                        {
    for(i=0;i<n;i++)                            a[i]=b[i]+1;
        c[i]=a[i]/2;                            c[i]=a[i]/2;
                                            }
```

b. Loop fission

Loop fission is the inverse process of loop fusion, and divides a loop into multiple loops to improve the spatial locality. If there is an existing dependency in a loop, loop fission will divide it into different loops to eliminate this dependency.

An example of loop fission:

```
[code snippet]//original loop:          [code snippet]//split loop:
    for(i=0;i<n;i++)                        for(i=0;i<n;i++)
    {                                       {
        a[i]=a[i]+b[i-1];                       b[i]=c[i-1]*x*y;
        b[i]=c[i-1]*x*y;                        c[i]=1/b[i];
        c[i]=1/b[i];                        }
        d[i]=sqrt(c[i]);                    for(i=0;i<n;i++)
    }                                           a[i]=a[i]+b[i-1];
                                            for(i=0;i<n;i++)
                                                d[i]=sqrt(c[i]);
```

c. Loop tiling

Loop tiling technology can tile a loop into multiple nested loops according to the data features. By tiling, the loop tries to finish processing one data sheet before starting another, which improves the temporal locality of memory access. The size of each tile is decided by cache size. An example of matrix multiplication earlier shown in this chapter shows the impact of tiling technology. We also have to deal with the "tail" for tiling: for(i=it;i<min(it+nb,n);i++).

An example of loop tiling:

```
[code snippet]//original loop:          [code snippet]//tiled loop:
    for(i=0;i<n;i++)                        for(it=0;it<n;it+=nb)
        for(j=0;j<m;j++)                        for(jt=0;jt<m;jt+=mb)
            x[i][j]=y[i]+z[j];                      for(i=it;it<min(it+nb,n);i++)
                                                        for(j=jt;jt<min(jt+mb,m);jt++)
                                                            x[i][j]=y[i]+z[j];
```

d. Loop exchange

Loop exchange means changing the order of nested loops for data access to improve the spatial locality of cycling data access.

An example of loop exchange:

```
[code snippet]//original loop:          [code snippet]//interchanged loop:
for(j=0; j<m; j++)                      for(i=0;i<n;i++)
    for(i=0; i<n; i++)                      for(j=0;j<m;j++)
        c[i][j] = a[i][j] + b[j][i] ;           c[i][j] = a[i][j] + b[j][i] ;
```

The basic idea is that changing the execution order of instructions through code transform can improve the cache hit rate. Whether the loop-based code transform improves the cache performance is decided by the locality. Some difficulties, such as the correction of codes, the equivalence of executing programs, and modeling the related data access consumption, do exist in code transform. There also exist some advances in code transform, such as not needing to change the program's locality to improve cache performance, no platform dependency, and no special hardware is required (there is also no need to redesign code transform with the MIC version update).

2. Data transform

Code transform changes the instructions' execution order, while data transform changes the data location to improve cache performance according to the spatial locality. The basic idea of data transform is to group the data that are used often together. Thus, when the cache fails, we can load the related cache and its neighborhood, so cache failure is reduced. Two common methods of data transform are given below:

a. Data location

The variable's address is given when compiling or running, which decides the data's address in memory and the cache address. For example, in the virtual indexed cache, the variable's address in cache is the modulus of the variable's address and the size of the data's cache. Thus, we can put the variable at a suitable place for the related cache location. If frequently accessed variables are relocated within the same cache, then the spatial locality can be improved. In practice, we need to judge the data's relationship in selecting the data to be put together, such as the methods of clustering and coloring, which have obvious effects on pointer-based data structure.

b. Array reorganization

In the field of scientific computing, large data is organized in the style of an array. Reorganization of the data can effectively improve the cache performance. Loops exist in such programs, which access the arrays by the index. Each iterative step of the loop reads data from the memory area of each array. If we reorganize each of these arrays into a structure-based array, then the data used in each iterative step is in the same structure, and reading this data just requires access to a continual area, improving the spatial locality and reducing cache failures.

8.2.5   Vectorization Optimization

8.2.5.1 What Is Vectorization?

The Intel compiler supports vectorization, which uses a vector processor unit to perform batch computation. The loop section can be vectorized by an extended instruction set such as MMX, SSE, SSE2, SSE3, SSSE3, AVX, or Knights Corner instructions, thus greatly improving the calculation speed.

MIC supports 512-bit wide Knights Corner instruction and 16*32-bit or 8*64-bit processing mode, and the width of vectorization is 8 or 16. The length of 16 single-precision floating-point data is 512 bits. The operation of single-precision floating-point data vectorization is shown in Fig. 8.9. For example, for vector add: $C[0-15]=$

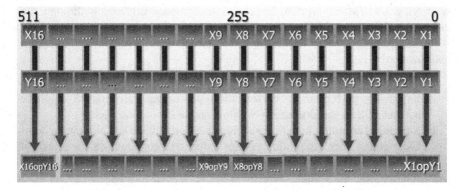

Fig. 8.9 Schematic diagram of single-precision floating-point data vectorization

A[0-15]+B[0-15] (A, B, and C are all float data), the calculation needs 16 add operations without vectorization. After vectorization, A, B, and C are put into the vector register. The previous add operation can be accomplished by only one operation. Therefore, vectorization can improve computing speed.

8.2.5.2 MIC Vectorization Optimization Strategy

MIC vectorization optimization mainly includes two methods: automatic vectorization and SIMD instruction optimization. Vectorization optimization is as follows:

1. Insert quotation automatic vectorization: without changing the original program structure, automatically vectorize by inserting a precompiled instruction (quotation).
2. Adjust program loop and insert quotation automatic vectorization: adjusting the original program structure to automatically vectorize by insertion.
3. Write SIMD instruction: SIMD instruction can achieve better performance compared with automatic vectorization. While the written SIMD instruction is dependent on hardware, and SIMD instruction has poor readability, it is certainly another option.

8.2.5.3 Automatic Vectorization

Automatic vectorization supplied by the Intel compiler is a function which automatically employs SIMD instruction. During data processing, the compiler automatically chooses instruction sets such as MMX, or Intel Streaming SIMD extension (Intel SSE, SSE2, SSE3, SSE4, AVX, and Knights Corner instructions), for the parallel processing of data. The function of automatic vectorization supplied by the compiler is an effective means for improving program performance. Automatic vectorization is supported well on the platforms IA-32 and x86-64. It can free the programmer to the maximum extent, and it not only shields the details of the underlying CPU/MIC, but also improves performance effectively through the underlying CPU/MIC SIMD parallelism. The MIC processor has a 512-bit wide vectorization processor, so automatic vectorization means a lot for MIC program optimization.

1. Advantages of automatic vectorization
    a. Improving performance: vectorization processing can process multiple batches of data simultaneously within a single instruction cycle.
    b. Simplified encoding by writing a single version of code and reducing assembling codes: less assembling coding reduces reprogramming for a special system, and the program is easily updated for the mainstream system without rewriting the assembling code.
2. What kind of the loop can be vectorized?
    a. For one single loop, if the compiler considers each statement as being independent or without dependence on the loop, the loop can then be vectorized. In other words, each statement should be implemented independently. The operation of writing and reading data must be neutral for iterations.

For example:

```
[code snippet]
for (int i=0; i<1000; i++)
{
    a[i] = b[i] * T + d[i] ;
    b[i] = (a[i] + b[i])/2;
    c = c + b[i];
}
```

which is equal to the following operations:

```
[code snippet]
for (int i=0; i<1000; i++) a[i] = b[i] * T + d[i] ;
for (int i=0; i<1000; i++) b[i] = (a[i] + b[i])/2;
for (int i=0; i<1000; i++) c = c + b[i];
```

Therefore, this loop can be vectorized.
Take another example:

```
[code snippet]
for (int i=1; i<1000; i++)
{
    a[i] = a[i-1] * b[i];
}
```

This loop can be vectorized, because a[i] in each iteration reads a[i-1] in the last iteration. This can be called cross-iteration data dependence or "flow dependence", and such a loop can't be vectorized by the compiler.

a. Vectorization can only act on the innermost loop: In a nested loop, the vector can only try to vectorize the innermost cycle. By checking the output information of each vector we can know whether the cycle is vectorized and why. If the key cycle that affects the performance isn't vectorized, you need to plan further, such as adjusting the order of the nested cycle.

b. The type of vectorization data should be consistent: variable types included in the statement that needs vectorization should be kept constant, such as trying to avoid single-precision and double-precision variables appearing in the same expression.

c. Compiler automatic vectorization method

2. Compiler vectorization options: For a MIC program, the default vectorization compile option is –vec, which means vectorization is opened by default, and you can close vectorization by adding –no-vec. The vectorization compiler can generate its own vectorization report, and such a function can be opened by the switch –vec-report. The detailed options are shown as Table 8.1:

Table 8.1 Vectorization report

-vec-report[n]	Functionality
n=0	Without displaying diagnosed result
n=1	Only displaying vectorized loops (default)
n=2	Displaying the vectorized and nonvectorized loops
n=3	Displaying the vectorized and nonvectorized loops, and also the dependent information
n=4	Only displaying nonvectorized loops
n=5	Displaying the nonvectorized loops, and also the dependent information

The usage of #pragma ivdep and restrict

In order to vectorize a loop which may include dependency, we usually add #pragma ivdep (ivdep, ignore, vector dependencies).

For example:

```
[code snippet]
1    void foo(int k)
2    {
3    #pragma ivdep
4      for(int j=0; j<1000; j++)
5      {
6        a[j] = b[j+k] * c[j];
7        b[j] = (a[j] + b[j])/2;
8        b[j] = b[j] * c[j];
9      }
10   }
```

When vectorizing this loop, the compiler considers array b to be dependent on the crossing iteration because of the usage of the variable k. If we know that k can't make data dependency, we can add #pragma ivdep to compile the guidance statement, ignoring the data dependence relationship, and try to vectorize it. The programmer must know how this dependency is generated, and make sure that it has no data dependency.

Make sure to use "#pragma vector always" to compile the guidance statement; this way of specified loop vectorization can avoid the case when some memory alignment operation cannot be vectorized, and using "#pragma simd" with the –simd option can force vectorizing in the innermost cycle. If using this method, the programmer needs to guarantee that the code is correct. There is a difference between the last statement and the previous two statements. The former two statements only suggest vectorizing, but the final result is decided by the compiler. However, the "#pragma simd" instruction forces the compiler to do the vectorization. If the compiler cannot vectorize the codes, it will give an error.

On MIC, using "#pragma vector aligned" can perform the vector alignment. However, with a guarantee of 64B memory alignment, that's align(64).

The loop with pointer may have dependency. To vectorize such a loop, we can use the keyword "restrict".

```
[code snippet]
1     void foo(float*restrict a, float*restrict b, float*restrict c)
2     {
3        for(int j=0; j<1000; j++)
4        {
5           a[j] = b[j] * c[j];
6           b[j] = (a[j] + b[j])/2;
7           b[j] = b[j] * c[j];
8        }
9     }
```

We should notice that adding the –restrict compiler option for using the keyword "restrict" is necessary, without which the compiler may account for the crossing iterations of array dependency. This is because the pointer is used to access data in the loop, so the compiler does not know whether the pointer is pointed to the same address (normally alias). Considering the security, we must stop such vectorization. The keyword "restrict" tells the compiler that the address the pointer indicates is limited, which can only be accessed by this pointer, or no alias.

3. Automatic vectorization optimization method

Reorder the nested loops

Automatic vectorization only happens in the innermost loop. However, the inner loop vectorization may not be perfect. We can reorder the nested loops for a better effect, including better continual access. In the following codes, the right side is better.

```
[code snippet]
for(j=0; j<N; j++)
#pragma ivdep
for( i=0; i<M; i++)
{
C[i][j] = A[i][j]+B[i][j];
}
```

```
[code snippet]
for( i=0; i<M; i++)
#pragma ivdep
for(j=0; j<N; j++)
{
C[i][j] = A[i][j]+B[i][j];
}
```

Split the loop

In some situations, besides the innermost loop, other loops are also time-consuming and not appropriate for vectorization. We can split these loops for more automatic vectorization. See the following codes.

Example 1:

```
[code snippet]
1    for(i=0; i<N; i++)
2    {
3        rand();
4        ...
5        ...
6    }
```

Assuming the loop does not have data dependency, the function rand cannot be vectorized; thus, the whole loop cannot be vectorized either. We can split the single loop into two loops to vectorize the second one (the time-consuming loop).

```
[code snippet]
1    for(i=0; i<N; i++)
2    {
3        rand();
4    }
5    for(i=0; i<N; i++) //automatic vectorization
6    {
7        ...
8        ...
9    }
```

Example 2:

```
[code snippet]
1    float s;
2    for(i=0; i<N; i++)
3    {
4        ...
5        s=...;
6        for(j=0; j<M; j++)// automatic vectorization
7        {
8                if(s>0)
9                {
10                   ...
11        }
12    }
13   }
```

By assuming a two-level loop without data dependency, not only is the inner loop for(j=0;j<M;j++) time-consuming, but the part for solving s cannot be vectorized. So, we can split the outer loop for better vectorization. The modified codes are:

```
[pseudo-code]
1    float s[16];
2    for(i=0; i<N; i+=16)
3    {
4          T=min(N-1,16);
5          for(k=0; k<T; k++) // automatic vectorization
6          {
7                ...
8                s[k]=...;
9          }
10         for(k=0; k<T; k++)
11         {
12               for(j=0; j<M; j++) // automatic vectorization
13               {
14                     if(s[k]>0)
15                     {
16                           ...
17                     }
18               }
19         }
20   }
```

Through splitting, we can vectorize more codes to obtain a better vectorization performance.

1. Parallelism and Vectorization

Because we can launch hundreds of threads on MIC, we need to guarantee enough threads. But in most applications, we often put the outer-most loop in parallel, so that fewer outer loops may impact the parallelism. Thus, we need to consider the balance of parallelism and vectorization.

Consider the following example:

```
[code snippet]
1    for(i=0;i<100;i++)
2    {
3          for(j=0;j<1024;j++)
4          {
5                ...
6          }
7    }
```

If we assume there is no data dependency for the two-level "for" in the code, the optimal number of threads to be deployed is around 200 for MIC parallelization.

The inner "for" can be vectorized automatically, while the outer "for" has only 100 iterations, which is not enough for MIC's best performance. For this situation, we can divide up the inner loop "for" in order to make the parallel codes satisfy the parallelism and vectorization. One splitting method is shown below:

```
[code snippet]
1      for(i1=0;i1<200;i1++)
2      {
3              for(j1=0; j1<512; j1++)
4              {
5                      i = i1/2;
6                      j = (i1%2)*512+j1;
7                      ...
8              }
9      }
```

Besides the functionality of MIC parallelism and automatic vectorization, this splitting method can also expand the number of cycles in the outer "for" loop. As long as the number of inner "for" loop is less than 16, this guarantees the vectorization effect.

Vectorization can improve performance, but for some applications it loses precision. Because vectorization uses the vector unit, i.e., for float-point computation, the processor sets 1 parity bit and 16 float-point operations to work simultaneously. However, the parity bit is limited, which causes the precision error. However, in most situations, this error can be tolerated.

8.2.5.4 SIMD Optimization

Single instruction multiple data (SIMD) can copy multiple operations into the vector registers. Obviously, SIMD has an advantage on performance with a synchronous way to execute the same operation for multiple data.

The vectorization level is shown in Fig. 8.10. With increasing levels, the programming language is lower, more complex, controls more, and the theoretical performance is higher. In the opposite way, the programming is easier, but without ideal performance.

The first generation of Intel MIC is KNC; Knights Corner instructions is the general name of SIMD instructions supported by KNC. It is similar to SSE and AVX. Using Knights Corner instructions, we can control vector computing in fine granularity.

Types of Knights Corner instructions:

1. Knights Corner instructions: detailed SIMD instructions, which is the subset of SIMD in the assembling instruction set.
2. Intrinsics of Knights Corner: the package of Knights Corner instructions (covers almost all the instructions) that can be thought of as the built-in C/C++ types.

Fig. 8.10 Vectorization
levels

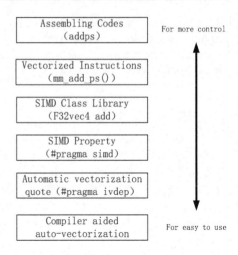

3. Knights Corner Class Libraries: in order to use Knights Corner instructions in a convenient way, the instructions are packaged so that the programmer can call SIMD instructions simply. It is a middle level between quotations and SIMD. Integer and float types are supported.

Next, we explain the difference between the three types. See Table 8.2.

Table 8.2 Vector Add in Knights Corner

Knights Corner Instructions	Intrinsics of Knights Corner	Knights Corner Class Libraries
__m512 a,b,c;	#include <immintrin.h>	#include <micvec.h>
__asm{ vloadd v0,b	. . .	. . .
vloadd v1,c	__M512 a,b,c;	F32vec16 a,b,c;
vaddps v0,v1	a = _mm512_add_ps(b,c);	a = b + c;
vstored a, v0 }	. . .	. . .

From the above example, we can see that using the class library is very simple, which is mostly similar to the scalar method (take the array as a variable) for vectorization. Using the instrinsics of Knights Corner directly is the common method, so we can compute the arrays through vectorize functions. The Instrinsics style is more complex compared with the class libraries, but with an improvement on performance. The inline assembling is most difficult to read and closest to the hardware level, thus with the highest performance and the most expensive programming work. In practice of programming with SIMD, we often use the intrinsics of Knights Corner.

Next, we give an example showing vector addition by using SIMD.

```
[code snippet]
1    #include <immintrin.h>
2    void foo(float *A, float *B, float *C, int N)
3    {
4    #ifdef __MIC__
5        __M512 _A, _B, _C;
6        for(int i=0; i<N; i+=16)
7        {
8            _A = _mm512_loadunpacklo_ps (_A, (void*)(&A[i]) );
9            _A = _mm512_loadunpackhi_ps (_A, (void*)(&A[i +16]) );
10           _B = _mm512_loadunpacklo_ps (_B, (void*)(&B[i]) );
11           _B = _mm512_loadunpackhi_ps (_B, (void*)(&B[i +16]) );
12           _C = _mm512_add_ps(_A, _B);
13           _mm512_packstorelo_ps ((void*)(&C[i]) , _C );
14           _mm512_packstorehi_ps ((void*)(&C[i +16]), _C );
15       }
16   #endif
17   }
```

SIMD is similar to an assembling language, with bad readability and dependency on hardware, which makes it difficult to migrate to other platforms. SIMD can be used selectively for compute-intensive codes.

8.2.6 Load Balance Optimization

8.2.6.1 What do we mean by Load Balance?

Load means the workload distribution among multiple tasks. Load balance denotes the even distribution of each task. For parallel computing, load balance puts the tasks into the resources to fully use the computer without idling or overloading. An effective parallel method can utilize the load balance perfectly. Load unbalance may cause a decrease in computing performance or bad extensibility. Thus, it's important to design the load balance for parallel computing, especially for the MIC, with its many cores.

In general, two strategies are used for load balance: static load balance and dynamic load balance. Static load balance needs to manually partition the working area into parallel parts with an equal workload. The dynamic load balance assigns tasks in the process of execution in a dynamic way. In practice, there exist a lot of problems that static load balance cannot solve, such as the computing workload, which is different in a loop. Normally, the performance of dynamic load balance is much better than the static situation, but this comes with more complex coding.

8.2.6.2 Load Balance Optimization for CPU/MIC Co-Computing

There are three levels of load balance for CPU/MIC co-computing:

1. load balance for threads/processes on computing devices (CPU or MIC)

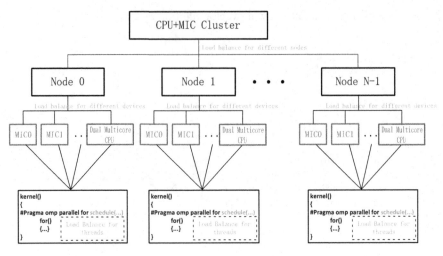

Fig. 8.11 Load balance for CPU/MIC co-computing

2. load balance between the CPU and MIC for one computing node
3. load balance between the computing nodes for clusters

Figure 8.11 shows the load balance levels for CPU/MIC co-computing.
1. Load balance on a computing device
There are three strategies for optimizing the load balance on a computing device
using OpenMP:

a. schedule(static [,chunk]): static scheduling method. Thread obtains the iterations
   of chunk with the style of poll. Without denoting chunk, it is dispatched on the
   average, which is the default method.
b. schedule(dynamic [,chunk]): dynamic scheduling method. The iteration is
   dispatched into threads dynamically. Without the parameter chunk, the iteration
   is operated one by one. With the parameter chunk, the iterative number for each
   thread is set to the value of chunk.
c. schedule(guided [,chunk]): guided scheduling method with self-guidance. In the
   beginning, each thread gets the larger iterative blocks, and then the blocks
   decrease. The size of the block decreases to the size of chunk exponentially.
   Without the parameter chunk, the iterative block size eventually decreases
   to one.

The available scopes of OpenMP scheduling strategies that can be used are
shown in Table 8.3.
2. Load balance between CPU/MIC
Because of the unequal performance of the CPU and MIC, the computing
workload on the CPU and MIC should be different. The best method of load balance
between the CPU and MIC is the dynamic load balance. We next introduce the load
balance between different devices for task schedule and data schedule.

Table 8.3 The available scopes of OpenMP scheduling strategies

Scheduling strategies	Available scopes
Static	Fixed amount of computation, with fixed iteration workload
Dynamic	Non-fixed amount of computation, without fixed iteration workload
Guided	Special dynamic scheduling strategy; the guided algorithm can reduce the workload

a. Task schedule: we can apply the dynamic load balance between the CPU and MIC for the applications of task schedule. For example, there are N tasks for two MIC cards in a node (three devices in all: one CPU, two MIC cards). Dynamic load balance needs each device to get a task for computation, and afterwards obtain the next task without waiting until the other devices are finished. This method only requires a main process to dispatch tasks for each computing process.
b. Data schedule: due to the single allocation of memory space on the device, we cannot apply dynamic load balance for the data schedule program. But the static load balance is not appropriate for heterogeneous computing. For some iteration applications, we can apply a learning style data scheduling method, such as the CPU and MIC doing one iteration for the same iteration, then compare the time as evidence for the data schedule.
3. Load balance between different nodes
   For different nodes, the load balance method is the same as the traditional one for the CPU cluster, for which the static and dynamic load balance are both appropriate.

8.2.6.3 Optimization of Thread Affinity

The load balance introduced in the previous sections assigns the tasks for different threads such that each thread runs almost at the same time. The threads run on the cores of the processor. Thus, in this section, we talk about how to schedule the threads on different cores running at the same time for load balance.

The thread affinity optimization covers the following situations:

1. Thread only knows the logic cores, so that a physical core using HT technology is also considered as two computing cores.
2. This kind of optimization is used for cache usage and physical overload problems, for which each is a constraint on the other.
3. It is used normally for situations in which not all of the physical cores can be utilized or there exists a locality problem.

So threads are dispatched on different logic cores according to different affinity settings. If there is data dependency between threads, then threads may utilize the cache on the same physical core for performance when the logic cores are on the same physical core. However, the computing work is on the same overload physical core while the other cores are fairly free. This situation still decreases the performance, especially when not all of the logical cores are working.

We need to configure the environmental variable KMP_AFFINITY for thread affinity optimization. Please refer to the OpenMP environmental variable setting in

Chap. 4 for a detailed explanation. Chapter 4 introduced the idea that the affinity optimization can be set by the scatter, compact, and balanced (MIC-only) modes.

The scatter mode dispatches the threads onto the light-load physical cores. However, because the neighboring threads may not be in the same physical core, with data being shared between the neighboring threads, the performance cannot be improved using cache. On MIC, the other cores' L2 cache is readable, but this may cause a reduction of free cache and slow speeds in reading from the other core.

The compact mode dispatches the threads on the logical cores in order, which tries to put the neighboring threads in the same physical core. With shared data between the neighboring threads, this mode can use cache as much as possible. If there is only a small difference between the neighboring threads, it will cause a serious load unbalance. But, if using a special task scheduling method, such as scheduling workload differently on odd and even threads respectively, an effective load balance can be achieved.

This balanced mode is unique for the MIC, though similar to the scatter mode. The difference is that the balanced mode considers a balance between the load and cache usage. It also tries to arrange the neighboring threads on the same physical core.

We should notice that these three modes are all static; thus, a special setting according to the actual workload in the execution process is necessary.

8.2.7   Extensibility of MIC Threads Optimization

8.2.7.1 What Does Extensibility of Threads Mean?

Extensibility means the performance improvement with the increase in threads under the same computing case. For example, the original program has four threads running on a quad-core CPU, and then due to the update of the CPU with eight cores, the program is also changed into eight threads. In this situation, the theoretical performance should be improved by a factor of two.

Please note:

1. The extensibility is considered that, with enough hardware resources, the number of threads should be less than or equal to the number of computing cores. Here, the computing cores mean logic cores, including the HT cores. But for some CPUs, the performance of HT may not be perfect; thus, sometimes we take the physical cores as the basis for the extensibility test. Even when each thread cannot fully utilize the hardware, each core running two or three threads can achieve better performance. But this situation does not exist for computing-intensive MIC programs. In practice, due to the hardware limitation, we must now discuss extensibility without considering computing cores.
2. In general, extensibility means the performance improvement after increasing the number of threads.
3. The ceiling of the speedup equals the increment of threads. When the threads number increases two times, then the performance is improved two times as well. This is known as a linear increment. But this ideal situation is hard to

achieve, especially with the growing number of threads. We need to notice the inflection point, or the point of performance degradation. Optimization should try to extend the inflection point after the use of computing cores. Assuming that the MIC card has 200 threads, we should try to make the performance increase linearly within 200 threads. Occasionally the increment is superlinear: the performance is improved by more than the thread growth. This situation mostly appears when the algorithm is changed, or with a better use of the cache.

8.2.7.2 How Do We Improve Extensibility for Many Threads ?

Our objective is to obtain linear speedup. Three aspects affect this objective, including:

1. Algorithm: the hotspot is partly dependent, incompletely parallelized, or just play a small part.
2. Parallelism: without a high degree of parallelism, adding threads is useless.
3. Hardware bottlenecks: because we take the computing cores as the ceiling for the number of threads, the computing cores cannot be the bottleneck. The bottleneck is normally the memory size, memory bandwidth, network bandwidth, etc.

It's a complex process to improve the extensibility, which is also the final objective of MIC program optimization. Thus, we should apply the utilization of various methods introduced in the previous sections to achieve this objective.

1. Algorithm: changing the algorithm is generally the fundamental method for many problems, including thread extensibility. We should reduce the thread dependency, utilizing an algorithm with high parallelism to achieve a theoretical linear speedup. Without such a theoretical linear speedup, we cannot obtain better performance with the limitation in practice. In addition, we should notice that the percentage of hotspots is not only in sequential codes. When the hotspots are parallelized, the running time is shortened, as is the hotspot percentage. Once the shortened scale reaches a degree (some percent on the CPU while tens of percent on the MIC), even the extensibility of the hotspot is good; for the whole program, however, the extensibility will be bad. This is not a common problem for CPU parallel optimization.
2. Parallelism: Please see Sect. 8.2.1 for parallelism optimization. One advantage is that if the original program is an MPI version running on a cluster, we can try to run one or multiple MPI processes on one node and run an OpenMP parallel subloop in the MPI inner process. This mixed MPI+OpenMP style may improve the performance significantly if the code is appropriate for this kind of mixed style (i.e., a good parallelism in the MPI inner process).
3. Hardware bottleneck: the bottleneck is often found in memory bandwidth, memory size, etc. Regarding the limitation of memory size, which causes the program to not be extended to more cores, see Sect. 8.2.2, Memory Management Optimization. Regarding the limitation of the memory bandwidth problem, see Sect. 8.2.4, Memory Access Optimization.

MIC Optimization Example: Matrix Multiplication

9

In this chapter, we introduce the programming model and optimization methods by using an example in matrix multiplication. The optimization covers computational methods, the communication between the CPU and the MIC, and the linkage between the CPU and the MIC.

Chapter Objectives.

- Learn how to implement matrix multiplication.
- Learn optimization methods of matrix multiplication on the MIC.
- Comprehend MIC optimization by specific examples.

9.1 Series Algorithm of Matrix Multiplication

For this example we have a matrix C=A*B, where A is an M*K matrix, B is a K*N matrix, and C is a M*N matrix. The main function structure of matrix multiplication is shown in Fig. 9.1.

E. Wang et al., *High-Performance Computing on the Intel® Xeon Phi™*,
DOI 10.1007/978-3-319-06486-4_9, © Springer International Publishing Switzerland 2014

```
[pseudo code]
int main(void)
{
    1.  //allocate and initialize the matrix A, B, and C
        // read the matrix A and B
        ...
    2.  //execute C=A*B on MIC card
    3.  //export matrix C
        //release matrix A, B and C
        ...
        return 0;
}
```

Fig. 9.1 Main function structure of matrix multiplication

The sequential algorithm of matrix multiplication is shown in Algorithm MatrixMul_V1, which contains three levels of loops. The inner loop iterates by variable k, in which a row of matrix A is multiplied by a column from matrix B. Then the product is an element of matrix C. In the two outer-level loops, for loops over every element in matrix C, i loops over the rows, and j loops over the columns.

Algorithm MatrixMul_V1

```
[code snippet]
    1   for(i=0;i<M;i++)
    2   {
    3       for(j=0;j<N;j++)
    4       {
    5           float sum = 0.0f;
    6           for(k=0;k<K;k++)
    7           {
    8               sum += A[i*K + k] * B[k*N + j];
    9           }
    10          C[I * N + j] = sum;
    11      }
    12  }
```

In this example, the size of matrix A, B, and C is 4096*4096. The sequential program consumes 312.83s on the Intel Xeon 5675 3.07GHz platform. (This program can be marked as P_baseline.) The analysis result of VTune is shown in Fig. 9.2, from which we can see that most of the time is consumed in the instruction

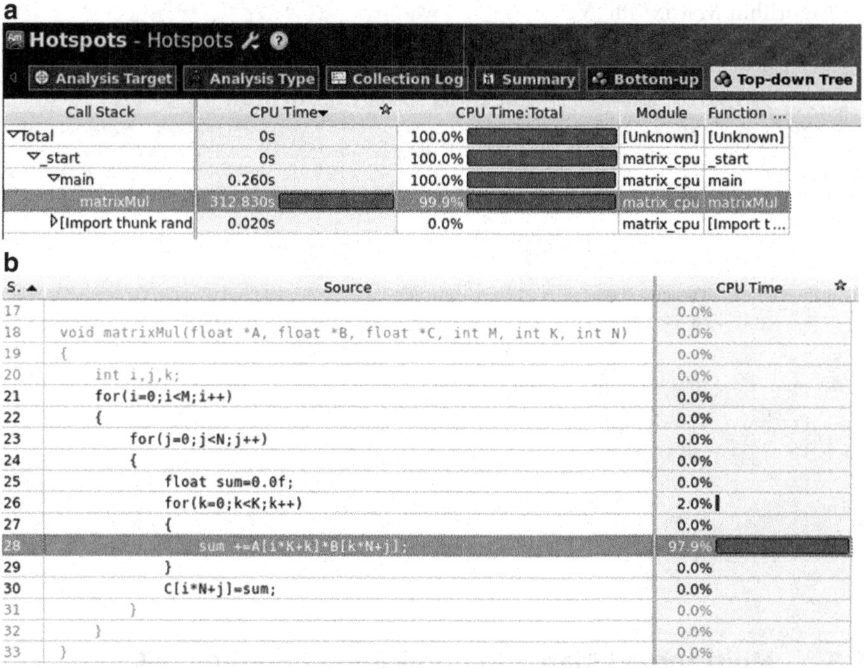

a

Hotspots - Hotspots						

Call Stack	CPU Time▾	☆	CPU Time:Total	Module	Function ...
▽Total	0s		100.0%	[Unknown]	[Unknown]
▽ _start	0s		100.0%	matrix_cpu	_start
▽main	0.260s		100.0%	matrix_cpu	main
matrixMul	312.830s		99.9%	matrix_cpu	matrixMul
▷[Import thunk rand]	0.020s		0.0%	matrix_cpu	[Import t...

b

S. ▲	Source	CPU Time ☆
17		0.0%
18	void matrixMul(float *A, float *B, float *C, int M, int K, int N)	0.0%
19	{	0.0%
20	int i,j,k;	0.0%
21	for(i=0;i<M;i++)	0.0%
22	{	0.0%
23	for(j=0;j<N;j++)	0.0%
24	{	0.0%
25	float sum=0.0f;	0.0%
26	for(k=0;k<K;k++)	2.0%
27	{	0.0%
28	sum +=A[i*K+k]*B[k*N+j];	97.9%
29	}	0.0%
30	C[i*N+j]=sum;	0.0%
31	}	0.0%
32	}	0.0%
33	}	0.0%

Fig. 9.2 Results of serial matrix multiplication in VTune

"sum += A[i*K + k] * B[k*N + j]". This instruction is in the third level of the loops, in which there is no dependence in the loops except for the innermost level.

9.2 Multi-thread Matrix Multiplication Based on OpenMP

According to the sequential matrix multiplication algorithm, we could implement the parallel version based on OpenMP. From the sequential code we could see, there is not any dependency in the two outer loops. Therefore the two outer loops can be readily parallelized by OpenMP. (The number of loops in the outer level must be greater than the number of threads.) In the Algorithm MatrixMul_V2, the variable THREAD_NUM is the number of threads. For example, on the two-channel Xeon 5675 3.07GHz platform with 6 cores in each CPU, the THREAD_NUM could be set to 24 (with Hyper-Threading switched on). In this situation, the program consumes 170.83s and the speedup is 1.83. This program can be designated with P_OMP.

Algorithm MatrixMul_V2

```
[code snippet]
1    #pragma omp parallel for private(j, k) num_threads(THREAD_NUM)
2    for(i=0;i<M;i++)
3    {
4        for(j=0;j<N;j++)
5        {
6            float sum = 0.0f;
7            #pragma ivdep
8            for(k=0;k<K;k++)
9            {
10               sum += A[i*K + k] * B[k*widthN + j];
11           }
12           C[i*N + j] = sum;
13       }
14   }
```

9.3   Multi-thread Matrix Multiplication Based on MIC

9.3.1   Basic Version

After creating the OpenMP version, we can begin to run the program on MIC with offload mode. Shown in Algorithm MatrixMul_V3.1, the directive "#pragma offload target(mic)" shows that the data will be offloaded to MIC. "In()" and "out ()" show the data transfer from the CPU to the MIC and the MIC to the CPU, respectively. "Length" is the length of data transmitted. The program is vectorized automatically by "#pragma". This version of program consumes 174s on the KNC platform (60 cores, 1.0GHz, 240threads), and the speedup is 1.80. This program can be marked as P_MIC_base.

Algorithm MatrixMul_V3.1

```
[code snippet]
1    #pragma offload target(mic)\
2    in(i, j, k, M, K, N)\
3    in(A: length(M*K))\
4    in(B: length(K*N))\
5    out(C: length(M*N))\
6    {
7          #pragma omp parallel for private(j, k) num_threads(THREAD_NUM)
8          for(i=0;i<M;i++)
9          {
10            for(j=0;j<N;j++)
11            {
12                float sum = 0.0f;
13                #pragma ivdep
14                for(k=0;k<K;k++)
15                {
16                    sum += A[i*K + k] * B[k*widthN + j];
17                }
18                C[i*N + j] = sum;
19            }
20        }
21   }
```

9.3.2   Vectorization Optimization

In Algorithm MatrixMul_V3.1, although one can employ automatic vectorization, because of the sum operation in the inner loop and the discontinuity of the array B access, the result which comes out is not so good. Instead, a better result can be achieved by interchanging the orders of loops. As shown in Algorithm MatrixMul_V3.2, the array B and C could be accessed continuously in the modified program. In array A, only one element is accessed in the inner loop.

The modified sequential version without vectorization consumes 53.07s on the CPU platform, while after vectorization, it is 25.00s. This version could be marked as P_baseline_vec, which runs 12.24 times faster than P_baseline. After the same optimization, the OpenMP version of this program consumes 4.53s, which is marked as P_OMP_vec. It runs 5.52 times faster than P_baseline_vec. The MIC version consumes 3.43s after optimization, and runs 7.92 times faster than P_baseline_vec. This program can be designated as P_MIC_vec.

Algorithm MatrixMul_V3.2

[code snippet]
```
1    #pragma offload target(mic)\
2    in(i, j, k, M, K, N)\
3    in(A: length(M*K))\
4    in(B: length(K*N))\
5    out(C: length(M*N))\
6    {
7        #pragma omp parallel for private(j, k) num_threads(THREAD_NUM)
8        for(i=0;i<M;i++)
9        {
10           for(k=0;k<K;j++)
11           {
12               float temp = 0.0f;
13               #pragma ivdep
14               for(j=0;j<N;k++)
15               {
16                   C[i*N + j] = temp * B[k*N + j];
17               }
18           }
19       }
20   }
```

9.3.3   SIMD Instruction Optimization

In order to improve on the performance of MIC-version programs even more, SIMD instructions could be applied. The MIC version with SIMD instructions is shown in Algorithm MatrixMul_V3.3. Matrix multiplication consumes 2.00s when SIMD instructions applied. This is marked as P_MIC_simd, which runs 12.5 times faster than P_baseline_vec, and 71.5% faster than P_MIC_vec. Although some programs could be accelerated by SIMD instructions, however, this makes the program more difficult to read and to understand. So the SIMD instructions are optional.

Algorithm MatrixMul_V3.3

```
[code snippet]
1      #pragma offload target(mic)\
2      in(i, j, k, M, K, N)\
3      in(A: length(M*K))\
4      in(B: length(K*N))\
5      out(C: length(M*N))\
6      {
7          #pragma omp parallel for private(j, k) num_threads(THREAD_NUM)
8          for(i=0;i<M;i++)
9          {
10             #ifdef __MIC__
11             __m512 _A, _B, _C;
12             for(k=0;k<K;j++)
13             {
14                 _A = _mm512_set_1to16_ps(A[i*K + k]);
15                 for(j=0;j<N/16;j+=16)
16                 {
17                     _B = _mm512_loadumpacklo_ps(_B, (void*)(&B[k*N + j]));
18                     _B = _mm512_loadumpackhi_ps(_B, (void*)(&B[k*N + j + 16]));
19                     _C = _mm512_loadumpacklo_ps(_B, (void*)(&C[i*N + j]));
20                     _C = _mm512_loadumpackhi_ps(_B, (void*)(&C[i*N + j + 16]));
21                     _C = _mm512_ad_ps(_C, _mm512_mul_ps(_A, _B));
22                     _mm512_packstorelo_ps((void*)(&C[i*N + j]), _C);
23                     _mm512_packstorehi_ps((void*)(&C[i*N + j + 16]), _C);
24                 }
25             }
26             #endif
27         }
28     }
```

The performance of the whole optimizations is shown in Fig. 9.3, and the speedup in Fig. 9.4.

9.3.4   Block Matrix Multiplication

For large matrix multiplication, we employ mostly block matrices, which benefits MIC optimization because:

1. Block matrix multiplication can make a better use of cache, increase the hit ratio, and then improve performance.
2. Block matrix multiplication can use the dual buffer and hide the communication between the CPU and the MIC by the MIC nocopy technique.

Program Version	Time (s)
P_baseline	312.83
P_OMP	170.83
P_MIC_base	174
P_baseline_vec	25
P_OMP_vec	4.53
P_MIC_vec	3.43
P_MIC_simd	2

Fig. 9.3 Performance of matrix multiplication

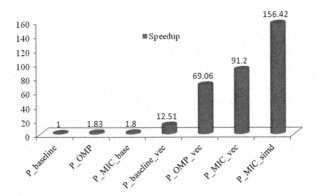

Fig. 9.4 Speedup of matrix multiplication

3. Because of the limited memory, block matrix multiplication con be applied in matrix multiplication of any scale.
4. Block matrix multiplication can create a good load balance between the CPU and the MIC by the means of allocating jobs dynamically for the CPU+MIC hybrid architecture.

We now introduce the optimized algorithm for large block matrix multiplication on the MIC.

9.3.4.1 Block Matrix Multiplication

Matrix multiplication can be denoted by $C_{m*n} = A_{m*k} * B_{k*n}$, which is shown in Fig. 9.5. There are three procedures in matrix multiplication:

Step 1: Partition the matrix in the direction i(for(i=0;i<M;i++)), which is shown in Fig. 9.6(a). The matrix A and C are partitioned by Mc.

Step 2: Based on Step 1, partition the matrix in the direction k(for(k=0;k<K;k++)), which is shown in Fig. 9.6(b), and the dimension of each submatrix is Kc.

Step 3: Based on Steps 1 and 2, partition the matrix in the direction j(for(j=0; i<N;j++)), which is shown in Fig. 9.6(c). The matrices B and C are partitioned by Nc.

The sequential algorithm of matrix multiplication is shown in the Algorithm MatrixMul_V4.1.

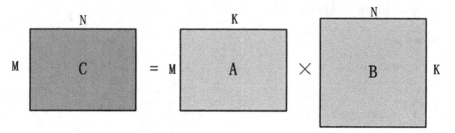

Fig. 9.5 Diagram of matrix multiplication

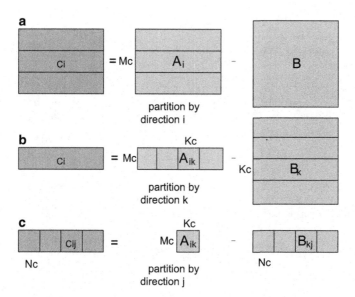

Fig. 9.6 Partition method of matrix multiplication

Algorithm MatrixMul_V4.1

[code snippet]

```
1    #define Mc 1024
2    #define Kc 1024

3    #define Nc 1024

4

5    void matrixMul(float *A, float *B, float *C, int M, int K, int N)

6    {

7        int i, j, k;

8        int ii, jj, kk;

9        int i_end, j_end, k_end;

10

11       i_end = Mc;

12       for(ii=0;ii<M;ii+=Mc)

13       {

14         if(Mc>M-ii)

15             i_end = M-ii;

16         k_end = Kc;

17         for(kk=0;kk<K;kk+=Kc)

18         {

19             If(Kc>K-kk)

20                 k_end = K-kk;

21             j_end = Nc;

22             for(jj=0;jj<N;jj+=Nc)

23             {

24                 if(Nc>N-jj)

25                     j_end = N-jj;

26

27                 for(i=ii;i<ii+i_end;i++)

28                 {

29                     for(j=jj;j<jj+j_end;j++)

30                     {

31                         float temp = 0;

32                         for(k=kk;k<kk+k_end;k++)

33                         {

34                             temp += A[i*K+k] * B[k*N + j];

35                         } //for(k=kk;k<kk+k_end;k++)

36                         C[i*N + j] += temp;

37                     } // for(j=jj;j<jj+j_end;j++)

38                 } // for(i=ii;i<ii+i_end;i++)

39             } // for(jj=0;jj<N;jj+=Nc)

40         } // for(kk=0;kk<K;kk+=Kc)

41     } // for(ii=0;ii<M;ii+=Mc)

42   }
```

$$
\begin{array}{|c|c|}\hline C_{00} & C_{01} \\\hline C_{10} & C_{11} \\\hline C_{20} & C_{21} \\\hline\end{array}
=
\begin{array}{|c|c|}\hline A_{00} & A_{01} \\\hline A_{10} & A_{11} \\\hline A_{20} & A_{21} \\\hline\end{array}
*
\begin{array}{|c|c|}\hline B_{00} & B_{01} \\\hline B_{10} & B_{11} \\\hline\end{array}
$$

Fig. 9.7 Example of block matrix multiplication

Partition by direction i	Partition by direction k	Partition by direction j	Computation process
ii=0	kk=0	jj=0	$C_{00}+=A_{00}*B_{00}$
		jj=1	$C_{01}+=A_{00}*B_{01}$
	kk=1	jj=0	$C_{00}+=A_{01}*B_{10}$
		jj=1	$C_{01}+=A_{01}*B_{11}$
ii=1	kk=0	jj=0	$C_{10}+=A_{10}*B_{00}$
		jj=1	$C_{11}+=A_{10}*B_{01}$
	kk=1	jj=0	$C_{10}+=A_{11}*B_{10}$
		jj=1	$C_{11}+=A_{11}*B_{11}$
ii=2	kk=0	jj=0	$C_{20}+=A_{20}*B_{00}$
		jj=1	$C_{21}+=A_{20}*B_{01}$
	kk=1	jj=0	$C_{20}+=A_{21}*B_{10}$
		jj=1	$C_{21}+=A_{21}*B_{11}$

Fig. 9.8 Process of block matrix multiplication

The process of block matrix multiplication is shown below by some examples (Fig. 9.7).

The computation process of block matrix multiplication is shown in Fig. 9.8.

9.3.4.2 Block Matrix Multiplication Based on the MIC

For matrix multiplication on the MIC, the cache usage and performance can be greatly enhanced by partitioning. The MIC version of block matrix multiplication is shown in Algorithm MatrixMul_V4.2. To test the impact of block matrix on performance, the matrix is set to 16384*16384. The primary time elapsed is 301.37s without partitioning, while the partitioned version only consumes 131.19s, which is 2.3 times faster. The same algorithm consumes 206.31s on the 2-channel, 8-core Xeon 5675 with 24 threads, while the MIC version is 1.57 times faster than OpenMP.

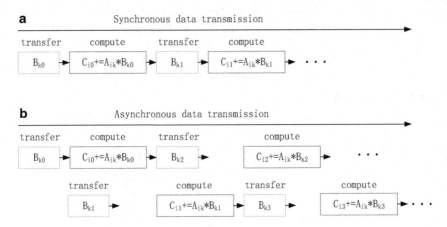

Fig. 9.9 Asynchronous block matrix multiplication

Algorithm MatrixMul_V4.2

9.3.4.3 Optimization of Asynchronous Matrix Multiplication

It is inefficient to transfer the data between PC memory and MIC memory by means of the PCI-E bus. Actually, the communication process between the CPU and the MIC can be hidden by using asynchronous computing, which is shown in the following example of matrix multiplication.

As is shown in the third step in Fig. 9.7), the blocks are transferred to MIC one at a time. This process on MIC could be shown in Fig. 9.9(a), and the asynchronous version is shown in Fig. 9.9(b), which introduces how to decrease the communication between the CPU and the MIC by asynchronization. The asynchronous version of block matrix multiplication is shown below in Algorithm MatrixMul_V4.3.

Algorithm MatrixMul_V4.3:

[code snippet]

```
1    #define Mc 1024
2    #define Kc 1024
3    #define Nc 1024
4
5    /*It should be declared globally when the pointer nocopy is used, and the keyword attribute should be
     added before that.*/
6    #pragma offload_attribute(push, target(mic))
7    float *Ac;
8    float *Bc0;
9    float *Bc1; /*Declare the double buffer space Bc0 and Bc1, which are used for asynchronous data
     transfer.*/
10   float *Cc;
11   #pragma offload_attribute(pop)
12
13   /*The functions called in offload must be defined by the keywords __attribute__ (( target (mic))) or
     __declspec( target (mic))*/
14   __attribute__ (( target (mic)))
15   void kernel(float *Ac, float *Bc, float *Cc, int i_end, int k_end, int j_end, int Kc, int Nc, int N, int ii, int jj, int
     THREAD_NUM)
16   {
17       int i, j, k;
18       #pragma omp parallel for private(i,j,k) num_threads(THREAD_NUM)
19       for(i=0; i<i_end; i++)
20       {
21           for(k=0; k<k_end; k++)
22           {
23               float temp = Ac[i*Kc +k];
24               #pragma ivdep
25               for(j=0; j<j_end; j++)
26               {
27                   Cc[(ii+i)*N +(jj+j)] += temp*Bc[k*Nc +j];
28               }//for(j=0; j<j_end; j++)
29           }//for(k=0; k<k_end; k++)
30       }//for(i=0;i<i_end;i++)
31   }
32
33   void matrixMul(float *A, float *B, float *C, int M, int K, int N, int THREAD_NUM)
34   {
35       int ii,jj,kk,jj0,jj1;
36       int i_end,j_end,k_end, j_end0, j_end1;
37
38       /*Allocate the partitioned space*/
39       Ac = (float *)malloc(sizeof(float)*Mc*Kc);
```

```
40      Bc0 = (float *)malloc(sizeof(float)*Kc*Nc);
41      Bc1 = (float *)malloc(sizeof(float)*Kc*Nc);
42      Cc = C;
43
44      /*Allocate space on MIC*/
45  #pragma offload target(mic:0) \
46      nocopy(Ac:length(Mc*Kc) alloc_if(1) free_if(0)) \
47      nocopy(Bc0:length(Kc*Nc) alloc_if(1) free_if(0)) \
48      nocopy(Bc1:length(Kc*Nc) alloc_if(1) free_if(0)) \
49      nocopy(Cc: length(M*N) alloc_if(1) free_if(0))
50      {
51      }
52
53      i_end=Mc;
54      for(ii=0;ii<M;ii+=Mc)
55      {
56          if(Mc>M-ii)
57              i_end=M-ii;
58          k_end=Kc;
59          for(kk=0;kk<K;kk+=Kc)
60          {
61              if(Kc>K-kk)
62                  k_end=K-kk;
63              for(i=0; i<i_end; i++)
64                  for(k=0;k<k_end; k++)
65                      Ac[i*Kc+k] = A[((ii+i)*K+(kk+k)];
66  #pragma offload target(mic:0) \
67              in(Ac:length(Mc*Kc) alloc_if(0) free_if(0))   //Transfer A(ii, kk)
68              {
69              }
70          j_end = Nc;
71          j_end0 = Nc;
72          j_end1 = Nc;
73          jj0=0;
74          if(Nc>N-jj0)
75              j_end0 = N-jj0;
76          for(k=0; k<k_end; k++)
77              for(j=0; j<j_end0; j++)
78                  Bc0[k*Nc+j] = B[(kk+k)*N+(jj0+j)];
79  #pragma offload_transfer target(mic:0) in(Bc0:length(Kc*Nc) alloc_if(0) free_if(0)) signal(Bc0) /*
    Asynchronous data transfer.*/
80          int js=0;
81          for(jj=0; jj<N; jj+=Nc, js++) //Partition B
82          {
```

```
83                    if(js%2==0)
84                    {
85                        jj1 = jj+Nc;
86                        if(jj1<N)
87                        {
88                            if(Nc>N-jj1)
89                                j_end1=N-jj1;
90                            for(k=0; k<k_end; k++)
91                                for(j=0; j<j_end1; j++)
92                                    Bc1[k*Nc+j] = B[(kk+k)*N+(jj1+j)];
93  #pragma offload_transfer target(mic:0) in(Bc1:length(Kc*Nc) alloc_if(0) free_if(0)) signal(Bc1) /*
    Asynchronous data transfer.*/
94                        }
95                        if(Nc>N-jj)
96                            j_end = N-jj;
97  #pragma offload target(mic:0) \
98                        in(i_end, k_end, j_end, Kc, N, Nc, ii, jj) \
99                        nocopy(Ac,Bc0,Cc) wait(Bc0)
100                       {
101                           kernel(Ac, Bc0, Cc, i_end, k_end, j_end, Kc, Nc, N, ii, jj, THREAD_NUM); //Call the
    kernel
102                       }
103                   }
104                   else
105                   {
106                       jj0 = jj+Nc;
107                       if(jj0<N)
108                       {
109                           if(Nc>N-jj0)
110                               j_end0=N-jj0;
111                           for(k=0; k<k_end; k++)
112                               for(j=0; j<j_end0; j++)
113                                   Bc0[k*Nc+j] = B[(kk+k)*N+(jj0+j)];
114  #pragma offload_transfer target(mic:0) in(Bc0:length(Kc*Nc) alloc_if(0) free_if(0)) signal(Bc0) /*
    Asynchronous data transfer.*/
115                       }
116                       if(Nc>N-jj)
117                           j_end = N-jj;
118  #pragma offload target(mic:0) \
119                       in(i_end, k_end, j_end, Kc, N, Nc, ii, jj) \
120                       nocopy(Ac,Bc1,Cc) wait(Bc1)
121                       {
122                           kernel(Ac, Bc1, Cc, i_end, k_end, j_end, Kc, Nc, N, ii, jj, THREAD_NUM); //Call the
    kernel
```

```
123                    }
124                }
125            }
126        }
127    }
128    /*Return results and release space on MIC */
129    #pragma offload target(mic:0) \
130        nocopy(Ac:length(Mc*Kc) alloc_if(0) free_if(1)) \
131        nocopy(Bc0:length(Kc*Nc) alloc_if(0) free_if(1)) \
132        nocopy(Bc1:length(Kc*Nc) alloc_if(0) free_if(1)) \
133        out(Cc:length(M*N) alloc_if(0) free_if(1))
134    {
135    }
136  }
```

Please note that 1) The core algorithm of block matrix multiplication is shown in lines 13–31. 2) The memory allocation on the MIC is shown in lines 44–51. 3) The asynchronous transfer is shown in lines 79–125.

3.1.) First the value of Bc0 (line 79) is transferred to prepare the data for asynchronization. The word offload_transfer is applied in asynchronization.

3.2.) The loop of partitioning matrix B is shown in line 81, in which the asynchronous transfer is applied.

3.3.) The procedure of asynchronous transfer is shown in lines 83–124:
  3.1.1) The "if" branch in lines 83–103 shows: The value Bc1 is transferred (line 93), which is used in core(line 101).
  3.2.2) The "else" branch in lines 104–124 shows: The value Bc0 is transferred line 114), which is used in core (line 122 4) The value of Cc is transferred back in lines 128–135, and the memory occupied on the MIC is now released.

9.3.4.4 Matrix Multiplication Based on CPU+MIC hybrid Computing

CPU and MIC are all based on x86 architecture, and the same optimization. So we can employ the paradigm of hybrid computing of CPU+MIC to greatly improve the computational performance on a CPU+MIC platform. Moreover, we can execute the same source code on both the CPU and the MIC. We then introduce below matrix multiplication based on two-way CPU + multi-MIC on a single node hybrid computing.

CPU+MIC hybrid computing can be achieved by MPI/OpenMP+offload mode. Here, matrix multiplication based on the single node CPU+MIC hybrid computing is achieved by MPI+offload mode, which is shown in Fig. 9.10. For this purpose, programmers can use the OpenMP+offload mode and multi-node version by themselves.

Matrix multiplication based on single-node CPU+MIC hybrid computing is shown in Algorithm MatrixMul_V4.3. All the CPU cores in the node can be

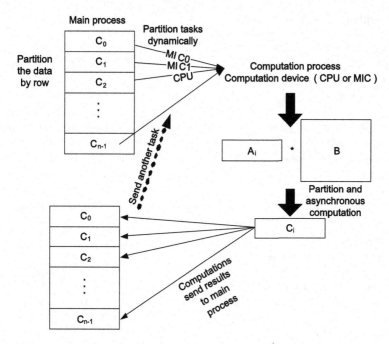

Fig. 9.10 Matrix multiplication based on single-node CPU+MIC hybrid computing

regarded as one device, and each MIC card could also be considered as a device in this node. If there are MIC_NUM MIC cards, the whole number of devices is MIC_NUM+1, and every device is controlled by an MPI process according to the process ID. The data allocation is achieved by the main process. In matrix multiplication, the data is allocated dynamically by dividing matrix C by rows into computing devices. Each amount of allocation is Mc*N. In other words, every device applies the data from main process and gets the results of Mc lines in matrix C. Then another set of data is applied until all the data have been multiplied.

Algorithm MatrixMul_V4.3

[code snippet]

```
1    #define Mc 1024
2    #define Kc 1024
3    #define Nc 1024
4
5    /* It should be declared globally when the pointer nocopy is used, and the keyword attribute should be
     added before that.*/
6    #pragma offload_attribute(push, target(mic))
7    float *Ak;
8    float *Bc0;
9    float *Bc1; /* Declare the double buffer space Bc0 and Bc1, which are used for asynchronous data
     transfer.*/
10   float *Cc;
11   #pragma offload_attribute(pop)
12
13   __attribute__ (( target (mic)))
14   void matrixMul (float *Ak, float *Bc, float *Cc, int i_end, int k_end, int j_end, int Kc, int Nc, int N, int jj, int
     THREAD_NUM)
15   {
16       int i, j, k;
17       #pragma omp parallel for private(i,j,k) num_threads(THREAD_NUM)
18       for(i=0;i<i_end;i++)
19       {
20           for(k=0; k<k_end; k++)
21           {
22               float temp = Ak[i*Kc +k];
23               #pragma ivdep
24               for(j=0; j<j_end; j++)
25               {
26                   Cc[i*N +(jj+j)] += temp*Bc[k*Nc +j];
27               }//for(j=0; j<j_end; j++)
28           }//for(k=0; k<k_end; k++)
29       }//for(i=0;i<i_end;i++)
30   }
31
32   int main( int argc, char *argv[] )
33   {
34       int THREAD_NUM_OMP = 1;
35       int THREAD_NUM_MIC = 1;
36       int M,K,N;
37       int myrank, root=0, totalrank;
38       MPI_Status status;
39       int MIC_NUM=2;
40       int deviceID=-1;
```

```
41      int nodeID=-1;
42      MPI_Init(&argc,&argv);
43      MPI_Comm_rank(MPI_COMM_WORLD,&myrank);
44      MPI_Comm_size(MPI_COMM_WORLD, &totalrank);
45
46      /*The main process controls the data partitioning, and allocates tasks dynamically according to the row of
        matrix C. The size of the matrix, which is allocated to the device each time, is Mc*N*/
47      if(myrank==root)
48      {
49          Initialize M, K, N, MIC_NUM, THREAD_NUM_OMP and THREAD_NUM_MIC;
50          MPI_Bcast(&MIC_NUM,1,MPI_INT,root,MPI_COMM_WORLD);
51          MPI_Bcast(&THREAD_NUM_OMP,1,MPI_INT,root,MPI_COMM_WORLD);
52          MPI_Bcast(&THREAD_NUM_MIC,1,MPI_INT,root,MPI_COMM_WORLD);
53          MPI_Bcast(&M,1,MPI_INT,root,MPI_COMM_WORLD);
54          MPI_Bcast(&K,1,MPI_INT,root,MPI_COMM_WORLD);
55          MPI_Bcast(&N,1,MPI_INT,root,MPI_COMM_WORLD);
56          float *A, *B, *C;
57          A =(float *)malloc(sizeof(float)*M*K);
58          B =(float *)malloc(sizeof(float)*K*N);
59          C =(float *)malloc(sizeof(float)*M*N);
60          int i,j,k;
61          int ii;
62          int processID;
63          int *flag = (int *)malloc(sizeof(int)*totalrank);
64
65          for(i=0;i<totalrank;i++) //Store the line number of the matrix C in each device.
66              flag[i] = -1;
67          //Initialize A and B;
68          MPI_Bcast(B, K*N, MPI_FLOAT, root, MPI_COMM_WORLD);
69
70          for(ii=0; ii<M; ii+=Mc) /*Allocate data dynamically and receive the results from each computation
        process.*/
71          {
72              MPI_Recv(&processID,   1,   MPI_INT,   MPI_ANY_SOURCE,   MPI_ANY_TAG,
        MPI_COMM_WORLD, &status); //Communicate with the computation processes.
73              if(flag[processID] != -1)
74                  MPI_Recv(CM+flag[processID]*N,  MIN(Mc,  M-flag[processID])*N,  MPI_FLOAT,
        processID, processID, MPI_COMM_WORLD, &status); //Receive the results from computation processes.
75              flag[processID] = ii;
76              MPI_Send(&ii, 1, MPI_INT, processID, processID, MPI_COMM_WORLD); /*Send line
        number.*/
77              MPI_Send(A+ii*K,   MIN(Mc,   M-ii)*K,   MPI_FLOAT,   processID,   processID,
        MPI_COMM_WORLD); //Send the partitioned matrix Aii.
78          }
```

```
79          for(i=1; i<totalrank; i++) //Notify all the computation processes that the tasks have been allocated.
80          {
81              MPI_Recv(&processID,   1,   MPI_INT,   MPI_ANY_SOURCE,   MPI_ANY_TAG,
            MPI_COMM_WORLD, &status); //Communicate with computation processes.
82              if(flag[processID] != -1)
83                  MPI_Recv(CM+flag[processID]*N,   MIN(Mc,   M-flag[processID])*N,   MPI_FLOAT,
            processID,  processID,  MPI_COMM_WORLD,  &status);  /*Receive  the  last  result  from  computation
            processes.*/
84              flag[processID] = -1;
85              MPI_Send(&ii,  1,  MPI_INT,  processID,  processID,  MPI_COMM_WORLD);  /*Notify  the
            computation processes that tasks have been completed.*/
86          }
87          free(...)
88      }
89      else //computation processes
90      {
91          float *B;
92          float *Ac;
93          int ii=-1,jj,kk,jj0,jj1;
94          int i, j, k;
95          int i_end,j_end,k_end, j_end0, j_end1;
96          int M, K, N;
97          MPI_Bcast(&MIC_NUM,1,MPI_INT,root,MPI_COMM_WORLD);
98          MPI_Bcast(&THREAD_NUM_OMP,1,MPI_INT,root,MPI_COMM_WORLD);
99          MPI_Bcast(&THREAD_NUM_MIC,1,MPI_INT,root,MPI_COMM_WORLD);
100         MPI_Bcast(&M,1,MPI_INT,root,MPI_COMM_WORLD);
101         MPI_Bcast(&K,1,MPI_INT,root,MPI_COMM_WORLD);
102         MPI_Bcast(&N,1,MPI_INT,root,MPI_COMM_WORLD);
103         deviceID = (myrank-1)%(MIC_NUM+1); //Compute the device(CPU or MIC) number on single
        node.
104
105         B =(float *)malloc(sizeof(float)*K*N);
106         Ac = (float *)malloc(sizeof(float)*Mc*K);
107         Ak = (float *)malloc(sizeof(float)*Mc*Kc);
108         Bc0 = (float *)malloc(sizeof(float)*Kc*Nc);
109         Bc1 = (float *)malloc(sizeof(float)*Kc*Nc);
110         Cc = (float *)malloc(sizeof(float)*Mc*N);
111
112         MPI_Bcast(B, K*N, MPI_FLOAT, root, MPI_COMM_WORLD);
113         if(deviceID<MIC_NUM) //The processes allocate space on MIC.
114         {
115             #pragma offload target(mic:deviceID) \
116                 nocopy(Ak:length(Mc*Kc) alloc_if(1) free_if(0)) \
117                 nocopy(Bc0:length(Kc*Nc) alloc_if(1) free_if(0)) \
```

```
118                    nocopy(Bc1:length(Kc*Nc) alloc_if(1) free_if(0)) \
119                    nocopy(Cc:length(Mc*N) alloc_if(1) free_if(0))
120                    {
121                    }
122              }
123       while(1)
124       {
125             MPI_Send(&myrank, 1, MPI_INT, 0, myrank, MPI_COMM_WORLD);/*Communicate with the
main process.*/
126            if(ii!=-1)
127                  MPI_Send(Cc, MIN(Mc, M-ii)*N, MPI_FLOAT, 0, myrank, MPI_COMM_WORLD);
/*Send results to the main process.*/
128                  MPI_Recv(&ii, 1, MPI_INT, 0, myrank, MPI_COMM_WORLD, &status); /*Receive line
numbers.*/
129            if(ii<M)
130            {
131                  MPI_Recv(Ac, MIN(Mc, M-ii)*K, MPI_FLOAT, 0, myrank, MPI_COMM_WORLD,
&status); /*Receive the partitioned matrix Ac from the main process.*/
132                  for(i=0;i<Mc*N;i++)
133                      Cc[i] = 0.0f;
134                  if(deviceID<MIC_NUM)
135                  {
136 #pragma offload target(mic: deviceID) in(Cc: length(Mc*N) alloc_if(0) free_if(0))
137                      {
138                      }
139                  }
140
141                  i_end=MIN(Mc,M-ii);
142                  k_end=Kc;
143                  for(kk=0;kk<K;kk+=Kc)
144                  {
145                      if(Kc>K-kk)
146                          k_end=K-kk;
147                      for(i=0; i<i_end; i++)
148                          for(k=0;k<k_end; k++)
149                              Ak[i*Kc+k] = Ac[i*K+(kk+k)];
150                      if(deviceID<MIC_NUM)
151                      {
152 #pragma offload target(mic: deviceID) in(Ak: length(Mc*Kc) alloc_if(0) free_if(0))
153                          {
154                          }
155                      }
156
157                      j_end = Nc;
```

```
158              j_end0 = Nc;
159              j_end1 = Nc;
160              jj0=0;
161              if(Nc>N-jj0)
162                  j_end0 = N-jj0;
163              for(k=0; k<k_end; k++)
164                  for(j=0; j<j_end0; j++)
165                      Bc0[k*Nc+j] = B[(kk+k)*N+(jj0+j)];
166                  if(deviceID<MIC_NUM)
167                  {
168  #pragma offload_transfer target(mic:deviceID) in(Bc0:length(Kc*Nc) alloc_if(0) free_if(0)) signal(Bc0)
     // Asynchronous data transfer.
169                  }
170
171              int js=0;
172              for(jj=0;jj<N;jj+=Nc,js++)
173              {
174                  if(js%2==0)
175                  {
176                      jj1 = jj+Nc;
177                      if(jj1<N)
178                      {
179                          if(Nc>N-jj1)
180                              j_end1=N-jj1;
181                          for(k=0; k<k_end; k++)
182                              for(j=0; j<j_end1; j++)
183                                  Bc1[k*Nc+j] = B[(kk+k)*N+(jj1+j)];
184                          if(deviceID<MIC_NUM)
185                          {
186  #pragma offload_transfer target(mic:deviceID) in(Bc1:length(Kc*Nc) alloc_if(0) free_if(0)) signal(Bc1)
187                          }
188                      }
189                  if(Nc>N-jj)
190                      j_end = N-jj;
191                  if(deviceID<MIC_NUM)
192                  {
193  #pragma offload target(mic: deviceID) \
194                          in(i_end, k_end, j_end, Kc, N, Nc, ii, jj) \
195                          nocopy(Ak,Bc0,Cc) wait(Bc0)
196                  {
197                      matrixMul(Ak, Bc0, Cc, i_end, k_end, j_end, Kc, Nc, N, jj,
     THREAD_NUM_MIC); //Call the kernel of MIC.
198                  }
199              }
```

```
200                     else
201                     {
202                             matrixMul(Ak,  Bc0,  Cc,  i_end,  k_end,  j_end,  Kc,  Nc,  N,  jj,
       THREAD_NUM_OMP); //Call the multi-thread kernel of CPU.
203                     }
204                 }
205                 else
206                 {
207                     jj0 = jj+Nc;
208                     if(jj0<N)
209                     {
210                         if(Nc>N-jj0)
211                             j_end0=N-jj0;
212                         for(k=0; k<k_end; k++)
213                             for(j=0; j<j_end0; j++)
214                                 Bc0[k*Nc+j] = B[(kk+k)*N+(jj0+j)];
215                         if(deviceID<MIC_NUM)
216                         {
217 #pragma  offload_transfer  target(mic:  deviceID)  in(Bc0:length(Kc*Nc)  alloc_if(0)  free_if(0))
       signal(Bc0)
218                         }
219                     }
220                     if(Nc>N-jj)
221                         j_end = N-jj;
222                     if(deviceID<MIC_NUM)
223                     {
224 #pragma offload target(mic: deviceID) \
225                             in(i_end, k_end, j_end, Kc, N, Nc, ii, jj) \
226                             nocopy(Ak,Bc1,Cc) wait(Bc1)
227                     {
228                             matrixMul(Ak,  Bc1,  Cc,  i_end,  k_end,  j_end,  Kc,  Nc,  N,  jj,
       THREAD_NUM_MIC);
229                     }
230                     }
231                     else
232                     {
233                             matrixMul(Ak,  Bc1,  Cc,  i_end,  k_end,  j_end,  Kc,  Nc,  N,  jj,
       THREAD_NUM_OMP);
234                     }
235
236                 }
237             }
238         }
239         if(deviceID<MIC_NUM)
```

```
240                   {
241  #pragma offload target(mic: deviceID) \
242                       out(Cc: length(Mc*N) alloc_if(0) free_if(0)) //Return the results from MIC.
243                   {
244                   }
245               }
246           }
247           else
248                   break; //Exit the loop.
249           }
250
251       if(deviceID<MIC_NUM)
252       {
253  #pragma offload target(mic: deviceID) \
254           nocopy(Ak: length(Mc*Kc) alloc_if(0) free_if(1)) \
255           nocopy(Bc0: length(Kc*Nc) alloc_if(0) free_if(1)) \
256           nocopy(Bc1: length(Kc*Nc) alloc_if(0) free_if(1)) \
257           nocopy(Cc: length(Mc*N) alloc_if(0) free_if(1))
258           {
259           } //Release the space on MIC.
260       }
261       free(...);
262   }
263   MPI_Finalize();
264   }
```

Note:
1. The main process is shown in lines 47–88.
First, the main process is required for broadcasting the initialized data to computing processes (lines 50–68). The main process allocates the data dynamically to the computation processes and receives the results from each computation process, which is shown in lines 70–78. The main process first receives the applications of computation processes (line 72), then checks if the computation processes have the results (line 73). If success (line 74), the main process will send the line number and block data to computation processes (line 76 and 77). Line 79–86 showss that the main process receives the results from computing processes for the last time and alert all the computation processes. Finally, the computation terminates.
2. The computation process is shown in lines 89–262.
   a. First the computation process receives initialized data from the main process (lines 97–102). Line 112 shows the value of matrix B is obtained. All the computation processes need the value of matrix B.
   b. Lines 113–122 show that the computation processes allocate memory on MIC, and Bc0, Bc1 are double buffer memory used for asynchronization operations.

c. The whole while loop continuously applies data from main process and computes Ci for computation processes.

d. At first the computation processes send the application from the main process, and then check for the results (line 126). They will send the results to the main process if there have been results, and then receive the line number ii (line 128). If ii<M, the computation hasn't completed. They will receive the Ac data from main process (line 131).

e. The computing procedures of computational processes are the same as the block matrix multiplication and asynchronous communication (lines 132–245).

Lines 251–260 show the applied memory on MIC is released.

Part III
Project Development

In this section, the development process of an HPC application based on MIC is introduced, and readers can learn how to apply MIC technology to an actual project. We show the process by means of two example projects.

After this section, readers will be able to put MIC technology into real practice, bringing the benefits to actual projects.

Developing HPC Applications Based on the MIC

10

Now we have learned the programming and optimization methods based on MIC, and we are able to write some MIC programs. However, how do we put MIC technology into practice, and how do we port the CPU programs onto the MIC platform? This chapter gives you all these answers.

Chapter Objectives.

- Learn hotspot testing
- Learn how to analyze programs
- Learn how to develop MIC programs and port CPUs program to the MIC platform

High-performance computing (HPC) applications development based on the MIC is similar to traditional programming procedures, including feasibility analysis, a general plan, then a detailed plan, coding, and testing. But there are some differences between MIC and CPU programs, because MIC programs are usually developed on parallel or sequential CPU programs. So it is important for MIC development to design, analyze, and test current programs. In this chapter we present our understanding and experience on actual MIC projects. Once you have a good understanding of this concept, you can learn readily how to develop and port MIC programs.

The development and porting procedure of MIC program is shown in Fig. 10.1. First, hotspot testing should be initiated to find the most time consuming part of the program. If there is no hotspot, there's no need to optimize. Otherwise, the hotspot should be analyzed further to confirm that it is appropriate to port to the MIC. In practice, if a new version has been developed, hotspot testing should be performed again to confirm whether there is more code that can be ported to the MIC.

E. Wang et al., *High-Performance Computing on the Intel® Xeon Phi™*,
DOI 10.1007/978-3-319-06486-4_10, © Springer International Publishing Switzerland 2014

Fig. 10.1 The
developmental procedure of a
representative MIC Program

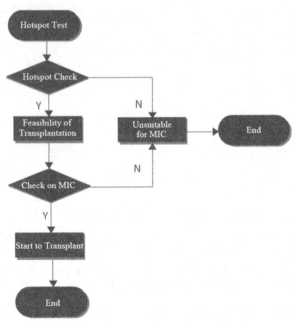

10.1   Hotspot Testing

10.1.1 Preparation

10.1.1.1 Theory Preparation

Readers who are interested in parallel computing should be familiar with the Amdahl's law, which is the most common criterion for testing the performance of parallel programs. Amdahl's law states that the speedup of a program using multiple processors in parallel computing is limited by the time needed for the sequential fraction of the program. So, we should be concerned with not only the optimization performance of hotspots but also the time proportion of hotspots to the whole. Otherwise, although the speedup of the hotspot parts of the program may be sufficiently high, the execution time as a whole may remain the same.

To test the optimization performance, we discuss the basics of speedup. Speedup is the ratio of the execution time of the sequential program to that of the parallel version. If the speedup is greater than 1, we could say that the optimization works. In scientific research, the test cases and testing environment should be referred to, but generally we need not be so strict. However, we should note that we cannot compare speedups performed under different testing environments or in different test cases.

The most commonly used speedup is:

$$\text{Speedup} = \left(T_s + T_{p'}\right) / \left(T_s + T_p\right)$$

where T_s is the executing time of the sequential part, T_p is the executing time of the parallelizable part, and $T_{p'}$ is the executing time of the parallelizable part after it has been parallelized.

10.1.1.2 Construction of Test Cases

In realistic situations, the part that needs to be accelerated is usually independent from the whole program. If the entire program is very long, we can extract the portion to be optimized. First, we analyze the code to be optimized, and the input and output data which should be presented. Then we prepare the value of input and output data that should be dumped from memory. Afterwards, we package the extracted code to an independent function. Finally, we can build a new main function and include the input and output data and the new function. The input data can ensure that the extracted portion runs successfully, and the output data can be used to check whether the result is correct. To dump the data from memory, we must save the input and output data to files in text or binary format (binary is preferred, because of its high precision), for example, we can use the function fwrite in C. These data can readily be used as input by the new program, and we can then check the result. Moreover, if the function to be optimized is called more than once, when extracted, the same procedure should be repeated again to ensure that they are identical.

10.1.2 Hotspot Locating and Testing

Before optimization, we should first find the hotspots by testing, and find the most time-consuming part of the program. Although some hotspots can be sighted directly, we also recommend that analysts do regular testing. Sometimes, the real hotspot may be different from what we anticipated.

Generally, there are two ways of testing hotspots:

1. Manually test by adding some timing functions.
2. Use hotspot analysis tools like VTune Amplifier XE.

In the first method, we should add timing functions before and after each section of the code we want to test, then we can get the executing time of this section. Using this method, we could get the execution time of every function or section of the program. The granularity must be considered carefully according to the actual project. If the granularity is too small (test every function or every line), this means a very heavy workload, and we may not get the exact results because of the short running time. Large granularity makes it very difficult to find the real hotspot (we may find that most of the code is full of hotspots!), and we cannot proceed any further. So typically we must test programs from large granularity to small. First, begin with large granularity to find the hotspot, and then partition it with small granularity, in which we test every line of code included by the section to avoid missing any potential hotspot.

When using the timing functions, we must choose the appropriate unit. For example, if the program needs a runtime of 10 h, it is not necessary to choose millisecond as the unit. A second would be fine in this case. Here we choose the function gettimeofday as our basic testing unit. However, although gettimeofday has high precision and is convenient to use, it also has restrictions.

The function gettimeofday is the system function in UNIX/Linux. There are newer versions like clock_gettime that can be substituted, but gettimeofday remains available. This function obtains the system time. We get time respectively in the beginning and the end of the program, and the difference between these times is the running time of the program (with units of microseconds). Before using this function, the code #include<sys/time.h> should be added in the beginning of the program. Moreover, most timing functions can obtain the actual time instead of the CPU time. But Fortran, the function CPU_TIME gets the CPU time (running time in the CPU). However, since the program will not use the CPU time when waiting for I/O, CPU time is meaningless to us. The actual time includes the wait time, which aligns with the reality of using the program. But because what we obtain is the actual running time, it means the system overhead could affect the performance. To get accurate performance results, it is better to run the test on idle nodes. An idle nodes mean there is no uncorrelated program running and little system service. If possible, any connection to the network should be avoided.

The use of gettimeofday is shown as follows:

```
[code snippet]
1   #include <sys/time.h>
2   int main()
3   {
4       struct timeval start,end;
5       gettimeofday(&start,NULL);/*The second parameter will get the time zone, which isn't needed
    here.*/
6       foo()//The running time of the functions that need to be timed mustn't be too short.
7       gettimeofday(&end,NULL); //If the return is 0, it is successful.
8       /*end.tv_sec–start.tv_sec is the time elapsed, whose unit is second. The data type time_t usually is
    same as long int.*/
9       /*( end.tv_usec–start.tv_usec is the time elapsed, whose unit is microsecond. Time elapsed
    suseconds_tusually is same as long int.*/
10      //1 second=10³ millisecond=10⁶ microsecond
11  }
```

There is a timing process system_clock in Fortran system functions. Please note that this function could return a different unit of value under a different system environment. If you want to use gettimeofday in Fortran, because the parameters are still all in structural format, it needs to be converted in Fortran, which is not convenient. If we package the gettimeofday, then it can be used in Fortran.

```
[code]
1    //gettime.c Compile by C compiler(not C++): icc –c gettime.c.
2    #include <sys/time.h>
3    #include <stdlib.h>
4    #include <stdio.h>
5    //Note that the function name has an underline, while there is no underline in Fortran codes.
6    //The results are different because of using different programming languages.
7    //The results are different because of using different programming languages.
8    void gettime_(long *time)
9    {
10       struct timeval tv;
11       gettimeofday(&tv,0);
12       *time= (long)(tv.tv_sec*1e6+tv.tv_usec);
13   }
```

In Fortran code, we can call this as given below:

```
[code snippet]
1    integer(8) :: starttime,endtime
2    call gettime(starttime)
3        !This section needs to be timed.
4    call gettime(endtime)
5    print *,'oneTime=',(endtime-starttime)/1.0e3,'ms'
```

When compiling the Fortran program, the file gettime.o should be added.

Another method to test for hotspots is to use tools such as Intel VTune Amplifier XE. We have discussed using VTune in previous chapters. The advantages of VTune can be summarized as follows:

1. Reduce the workload, since you do not need to add timing functions manually.
2. The results are clear without missing anything essential; they are also easy to read.
3. Analyze conveniently. This tool can provide some performance analysis automatically.

However, adding timing functions is also a good idea for programmers. It is better to use timing functions instead of testing software in the following situations:

1. Cost: most testing software is commercially available (except gprof), which can be costly.
2. Performance disturbance: testing software must sample from the original program, which could affect the performance. Although timing functions may take some time to run, they affect the program less because they are called upon only a few times.
3. Code section: if there is no obvious function called or you need to only test part of the program, it is convenient to use timing functions.
4. Functions are complex: when many system functions are called, like disk I/O, some testing software cannot accurately determine the running time.

5. Easy to use: only add some timing and output functions, instead of learning the large testing software and its use.

So, we can choose the hotspot testing method we use according to what makes sense for different projects. Or, we can combine them to accurately locate hotspots.

We must also note the selection of test cases matters. Different test case scales can have a huge influence on the distribution of hotspots. In particular, when the test case is small and only runs for several seconds, it is most likely that system overhead occupies the majority of the running time. When the scales of test cases change, the running time of the different parts of the program may vary accordingly. Usually we choose large-scale cases to test the demands of large-scale simulations and determine future requirements.

10.2   Program Analysis

10.2.1 Analysis of Program Port Modes

Before porting the program, determining which mode is adapted to the program should be analyzed first. There are three application modes in MIC: CPU-hosted, MIC-hosted, and CPU+MIC symmetrically hosted. The CPU-only mode and the MIC-only mode are included in the first two modes. The MIC-hosted mode requires the algorithm to be highly parallelized, and the data needs to be transferred there manually. So its application is rather limited; the same is true of the symmetric mode. If we consider the CPU as host, the parallel computing is allocated to the MIC. This mode may be more efficient, and is extensively used. In this section, we apply the offload mode, in which the CPU is considered as host and the MIC as the coprocessor.

10.2.2 Analysis of Size of the Computation

The size of the application makes a big difference on the program architecture. So it is important to choose the program architecture by analyzing the size of the application. Usually the scale or size can be divided into these versions: a single MIC in one machine, multi-MIC in one machine (or CPU+MIC hybrid computing, they are almost the same), and MPI single-MIC in multiple machines (one MIC in one machine) and MPI multi-machine hybrid computing. There are some applications in which one MIC could only be used by one process. In this way, if there are two MIC cards in one node and three MPI processes in an application, these three processes run on the two MIC and CPU, respectively. But please note that such a program is no different from the MPI single-MIC in multiple machines.

How can we confirm the computational size? First, we should consider the original program size. For example, if the original program is an MPI multi-machine version, usually the optimization is designed as MPI multi-machine. If the original program is designed as single machine version, no matter whether it has a single thread or is multi-threaded, we can optimize the design as a single-MIC one-machine version or multi-MIC in one machine version. If the program needs to be ported onto a MIC platform, usually we first consider writing the single-MIC version, then according to the scale of computation, we can tell whether the multi-MIC or hybrid computing version is needed or not.

Although the parallel mode greatly influences the running time, time is not the only factor. If the original program is based on MPI, because of its high performance, the execution time is acceptable to run in single-node mode after porting to the MIC. But we could get better performance if we make more use of MIC in cases involving larger program sizes.

10.2.3 Characteristic Analysis

We next discuss the two cases of hotspots: concentrated and dispersed.

10.2.3.1 Concentrated Hotspots
The VTune hotspot analysis of the Computational Fluid Dynamics program achieved by the Inspur Group and Northwestern Polytechnical University (Fig. 10.2) shows us that 99.1% of the hotspots occurred on the launch_LBCollProp function. So, the optimization of this function could bring a great deal of performance improvement. This improvement is in the interest of programmers, because falls under the 80-20 rule, in which 80% of the running time is expended by 20% of the code. Therefore we can make good progress by optimizing only one function.

In this case, we port the hotspot function to the MIC, and the rest of the program stays behind on the CPU.

10.2.3.2 Dispersed Hotspots
1. Every subroutine is a hotspot in a large program
    The VTune hotspot analysis of a project shared by Inspur Group and an institute is shown in Fig. 10.3. This program is based on MPI clusters, but it is sequential in each MPI node.
    The function solver_grapes called by main function is the hotspot in this program. In the functions called by solver_grapes, the first six include 90.5% of the hotspots. But the hottest function of these only accounts for 38.3% of the time loss. Even if we could optimize it to 0 s (although that's impossible), the running time of the whole program remains at 60% of its original, and the speedup is only 1.67. Moreover, there are still many functions called in the most time-consuming function. In this case, although we can optimize each function separately, this will consume a lot of time and cost. However, if the program is based on parallel architecture (like this program based on MPC), we could consider porting the whole program to the MIC or moving some MPI processes to the MIC. In this way, although we cannot take complete advantage of the MIC, some workload can be shared by the MIC; sometimes this is

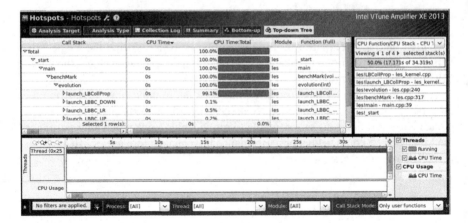

Fig. 10.2 Concentrated hotspots

Fig. 10.3 Dispersed hotspots

warranted. But note that using this mode will consume much more system resources (like memory), so task allocation must seriously be considered.

2. Multiple Hotspots

The VTune hotspot analysis of the image reconstruction program created by the Inspur Group and the Institute of Biophysics, Chinese Academy of Sciences is shown in Fig. 10.4. The hotspot of this program is in several functions instead of in just one function, and these functions with hotspots are very simple and are mostly basic functions. Optimization in this case is more practical and feasible, although we must still optimize several functions to improve the performance.

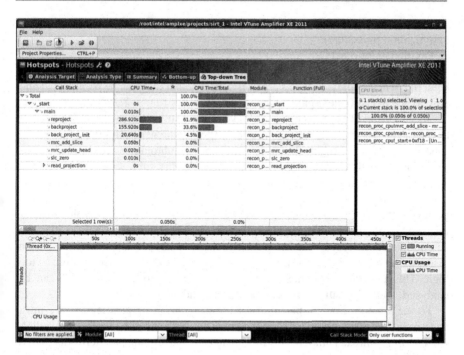

Fig. 10.4 Multiple hotspots

From this figure we can see that the two most time-consuming functions consume 61.9% and 33.6% of the time, respectively. So in theory, if these two functions are optimized, good performance can be gained. But in porting, we found that if only these two functions are optimized, but another function, which accounts for 4.5% of the time, remains in the CPU, much more data transfer has to be carried out between the CPU and MIC. This has a definite negative influence on the performance. Thus, we must port all three hotspot functions to derive the best performance in this program.

The difference between these two kinds of dispersed hotspots lies in the code complexity. One type is a large number of hotspot functions; the other is the large scale of each hotspot function. It is difficult to optimize the whole program if there is high complexity in the code. Otherwise, all the hotspots can be optimized, and the performance of the entire program will be improved.

10.2.4 Parallel Analysis of Hotspots

It is very difficult to tell whether hotspots are concentrated or not and confirm the hotspots by their granularity. We can only proceed with the parallel optimization combined with parallelism. The hotspots can be concentrated in only one function in spite of it consisting of short code. If this function cannot be parallelized, there remains nothing else that can be done about it (of course, we can choose other optimization procedure).

Parallelism or others?

Locating the hotspots is the primary step. There is so much more to consider here, like whether these hotspots are useful, or whether they can be parallelized and how to achieve sufficient parallelism.

The selection of hotspots can be achieved by the following method. The hotspot covered by the main function must be 100%, and generally there are some subroutines in the main function. And these subroutines could call other subroutines. In this circumstance, we can analyze the subroutine called directly by main function, and confirm whether this subroutine consumes most of the entire time and whether it can be called in parallel mode. Then, we should consider porting. Otherwise we should check whether the subroutines called by this subroutine can be called in parallel, one by one. There is no need at all to analyze all of the subroutines. Usually, only the hotspot functions are analyzed. But if the hotspots are dispersed, even if only the hotspots are parallelized, the cost of porting would be too high. This is clearly meaningless. So only the relatively large hotspots can be analyzed, and the small hotspots are usually abandoned.

To see whether there is potential for the program to be parallelized, we must check whether there are do loops in the program. Except for the recessive parallelism like multi-thread parallelized by pThread and coarse-grain parallelization based on MPI, there are loops in most parallel programs. After the loops are found, the iteration and the running time should be considered; usually large iteration times and long running time are preferred. We should also check if there is data dependence among iterations. Usually, to make a code section a good candidate for parallelization, the code section should meet these criteria: (1) there is loop; (2) there are many iterations (the number of iterations should be two to three times the number of cores, and the more the better).

Moreover, the memory on the MIC card should be considered. The same problem exists in the CPU too, but usually there is enough memory on the CPU platform, so it is not important. In parallel computing, every thread has its own memory. If there are few threads, or the memory owned by each thread is small, then there is no problem. However, if every thread takes a large amount of memory, or if there are plenty of threads, because there is less memory on the MIC than on the host, we must be careful.

10.2.4.1 Coarse-Grain Parallelization: MPI Level

Applications with few iterations, for which every iteration costs a lot of time, and for which there is little or no data transfer among iterations are appropriate for MPI-level parallelization. The main overhead is the network transfer. Because of the iterations among different nodes, we would prefer that more system resource (like memory) is required by a single iteration.

Coarse-grain parallelization is used for clearly divided tasks divided with little data transfer between tasks. The more time a single task consumes the better the performance we will obtain. This performance is usually limited by the slow speed of the network transfer among MPI processes. So if we choose MPI to parallelize our task, the speedup gained by using MPI must offset the latency produced by network transfer.

In another case, if every task requires many system resources, and there is more than one node available, because MPI supports non-shared memory clusters, we can easily allocate the tasks to different nodes to simultaneously meet the requirements of resources and speed.

10.2.4.2 Fine-Grain Parallelization

Fine-grain parallelization is usually based on iterations. Because not all the loops can easily be parallelized and not every parallelized loop can possibly improve performance, we must consider fine-grain parallization by our actual conditions or through experiments.

Generally, the parallelization of iterations should follow these criteria:

1. Feasibility

    After parallelization, correct results are the most important. To check whether the iteration could be parallelized, we need to see if there is data dependency among the various iterations.
2. Concurrency

    If it is feasible to do, concurrency is the key to performance improvement. Concurrency should not be too small; according to our experience, two to three times more than the number of cores is appropriate. For example, in a four-core node, in which the hyper-threading technology is closed, the iteration shouldn't be less than 8. While in MIC, using 50 cores for example, the iteration should not be less than 400. In this way, system resources are sufficient. Otherwise, the performance declines. Generally speaking, the greater number of iterations, the better the performance you can achieve.
3. Others

    Data conflict always gives rise to performance declines. The same memory is used by different iterations, and we usually encounter data conflicts when accumulation or reduction proceeds. We can avoid this by setting the variables private in different iterations, but there must be some sequential execution phases after accumulation or reduction. Although the parallel algorithm could improve performance, the final performance is actually determined by the time ratio of the sequential section to the whole program.

    The time consumed by a single iteration also plays an important part in the performance. After parallelization, there must be some extra overhead added. The gain brought by the parallelization may not offset this overhead in the case of too few iterations. Therefore developers must consider this problem seriously.

    There are two solutions in parallelizing the fine-grain loop: packaging (like OpenMP, Cilk) and threading (like pThread). It is convenient to use OpenMP, but we can get more control on the program by using pThread. Generally, OpenMP includes most of the functions we need.

    The programs ported to MIC are usually fine-grain. Even so, we could use MPI to achieve parallelization. Programs ported to MIC usually have these characteristics: high concurrency, little resources used, small workload in single thread, independent program sections, and little data transfer.

There is no contradiction between fine-grain and coarse-grain parallelization. In large projects, these two styles are always combined. In the following, we introduce some solutions in which the coarse-grain parallelization is carried out in the high level and the fine-grain parallelization in the bottom level. These solutions combine the advantages of these two styles and are appropriate for large projects.

10.2.5 Vectorization Analysis

A single MIC core is not faster than a, CPU, but the width of vectorization unit is 512 bit while the newest CPU architecture (Intel Ivy Bridge) is 256 bit. Vectorization is the most important method to obtain better performance.

Vectorization advances many data in the array by just one instruction. When the 512-bit vectorization unit advances the array A + B, both of which are float arrays, $512/32 = 16$ elements in the array can be advanced simultaneously.

We should find out which sections should be vectorized (loop and array) and which sections have been vectorized by the compiler automatically. Usually we could get this information from the compiler by adding the compiling option -vec-report3. Then we could confirm the ratio of the vectorized sections and see whether they are hotspots, and then determine how to vectorize the other sections.

If the hotspots have not been vectorized and there is no possibility to vectorize them, we must consider whether the application is appropriate for MIC. Then if the parallelism is not sufficiently high level, maybe the application should be ported to MIC. However, if the application has good parallel attributes, we can also obtain good performance by using a large number of threads. In this case, the application is still appropriate to be ported to MIC but may have some risks.

10.2.6 MIC Memory Analysis

1. Amount of Memory

Usually the memory on MIC is less than that on the host. There's enough memory and virtual memory for applications on the CPU because these applications are developed according to current amounts of memory. But after the application is ported to the MIC, the memory must be considered. Moreover, because of so many threads, no matter how much memory each thread needs, the whole amount of memory needed may be quite large. The memory needed by the application on the MIC may be much more than that on the CPU. So when developers begin to program on MIC, the amount of memory must first be considered.

Before porting to MIC, the amount of memory needed by the program and the memory associated to threads must be considered together. If the program itself requires a large amount of memory, we could consider whether the program can be divided to reduce the memory requirement. But when the number of threads increased, the memory required also increased. We can possibly consider canceling or combining some variables to reduce the memory requirements. Otherwise the concurrent number of threads will be seriously limited. Moreover, if possible, multi-MIC or CPU+MIC are good solutions.

2. Memory Bandwidth

   The bandwidth of each thread is less than the host. If a large amount of memory access occurs when an application is running, the performance will be terrible. In this case we can try to reduce the memory access or change the program structure to make sufficient use of the cache and finally to reduce the memory access.

10.2.7 Program Analysis Summary

If the program is appropriate to be ported to MIC, many factors should be considered. These include whether the program should be ported totally, or only a module, and whether the algorithm should be altered or not. Unless a parallel algorithm is required, other factors are not that important. Good performance can only be obtained if most of the requirements are met. For example, although the parallelism is not high enough, if the other requirements are met and perfect, then we could also get acceptable results, although the computation capability has not been used sufficiently.

In other words, the performance expectations and the program characteristics should be considered together. If the MIC only takes part of the computation, the program characteristics need not align to the MIC. But if we expect to get peak performance from the MIC, the requirements above should be met completely.

10.3 MIC Program Development

Once we confirm that the program can be parallelized and it is appropriate to run on MIC, then we can start the porting.

First, we must confirm the modules to be ported, which are usually hotspot functions or loops. Because there is data transfer between the host and the MIC, some codes around the hotspots should also be moved to MIC to reduce it. Because OpenMP is widely used, the example based on OpenMP on MIC is introduced. This is also appropriate for pThread-based programs.

The usual procedure for porting and developing on the MIC is shown in Fig. 10.5.

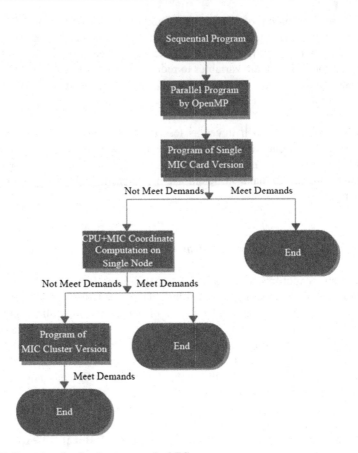

Fig. 10.5 Procedure for development on the MIC

1. Rewrite the program to the OpenMP version.
2. Optimize the OpenMP version appropriately (not all the optimizations on the CPU are appropriate for MIC, so we need not optimize to an extreme.)
3. Port the OpenMP version to the single MIC card.
4. Optimize the program according to the execution situation on the MIC.
5. If the MIC version of program has met the requirements, stop it. Otherwise, go on the next step.
6. Parallelization on the MIC cluster; include the MIC cluster computation and the CPU+MIC heterogeneous computation.

10.3.1 OpenMP Parallelism Based on the CPU

Before porting to the MIC, the sequential programs should be parallelized by OpenMP. Then the performance of the current program can be referred to when gauging the performance on the MIC. If the cores are increased, but the speedup remains the same, the current program should be optimized more to make sufficient use of the multi-core. But if the speedup still cannot reach our expectations, we must consider whether this program is appropriate for the MIC. In addition, if the speedup of the program parallelized by OpenMP is good enough, we can usually also expect to reach good performance on the MIC.

10.3.2 Thread Extension Based on MIC

If the OpenMP version of the program has been developed, we can begin to port it to MIC. The process is not complicated, and we can confirm the input and output parameters and the functions running on the MIC. That is enough.

However, there are some differences in computing between the CPU and the MIC. Because there are more concurrent threads on the MIC and the computing capability of a single core of MIC is not as powerful as the CPU, the low thread expansibility may hide the advantages of MIC, and the performance on the MIC may be worse than on the CPU. Therefore the thread expansibility must be optimized before porting to the MIC. Usually for good thread expansibility in OpenMP versions of programs on the CPU, we can get better performance on the MIC.

10.3.3 Coordination Parallelism Based on Single-Node CPU+MIC Mode

After the program has been ported to the MIC, the CPU only controls the whole process but does not take part in the computation. Sometimes the CPU may be left idle for a long time, which is very wasteful. Compared to the MIC, the CPU has the advantages of more powerful computing capability and no data transfer. So if the CPU and MIC can coordinate, we will get better performance.

In this case, we usually use pThread to control the CPU and the MIC respectively, both of which work well in OpenMP. The codes are shown below. In addition, because the computing capability of the MIC is different from that of the CPU, the tasks must be allocated equally; otherwise, one of the devices may not be utilized sufficiently.

```
[pseudo code]
1    #pragma omp parallel for
2    for(index=0;index<deviceNumber;++index)
3    {
4    if(index==0)
5       {
6              //Functions run on CPU.
7       }
8    else
9       {
10             //Functions run on MIC.
11      }
12   }
```

There is another situation in CPU+MIC coordinated computation. The CPU or the MIC in the node is regarded as on each device, respectively. When using MPI, every device is controlled by one MPI process. The MPI process controlling one MIC can also control another MIC by offload mode. The MPI process controlling the CPU is similar to the original CPU program (the task partition method may be changed). The MPI process controlling the MIC is similar to the original MIC program (again, the task partition method should be changed). It is advantageous that the cooperative computation mode can adapt different types of clusters, including different numbers of MICs, perfectly. But the overhead may be increased because of the large number of MPI processes.

10.3.4 MIC Cluster Parallelism

When one MIC node cannot meet this requirement, we must consider using clusters by the coarse-grain and fine-grain combined mode introduced above. There are two types of architecture in today's clusters: MIC nodes by themselves and MIC nodes + other CPU nodes.

10.3.4.1 Cluster of MIC Nodes

A cluster of MIC node means there are MIC cards in every node in the cluster. In this circumstance, MPI should be applied to both transfer data among nodes and to partition the tasks. To adapt different hardware and applications, dynamic partition mode is usually adopted. One process could be used as the main process to control data broadcasting; the peer to peer (P2P) mode could also be adopted. The program in the computing process usually can be based on a single MIC version or CPU+MIC version, and the input and output should be adjusted to the new program structure.

10.3.4.2 Cluster of MIC Nodes + CPU Nodes

A cluster of MIC nodes + CPU nodes means in the cluster there are not only the nodes with MIC cards installed, but also the traditional CPU nodes. Usually we can get this type of cluster when we add some MIC nodes to the current CPU nodes. In this circumstance, there's almost no difference in programming compared to the MIC version. Even the "offload" words need not be revised, because the program can run on the CPU when the driver cannot detect any MIC presence. But usually we need to set different number of threads to the CPU and the MIC to get better performance. The tasks should be allocated according to the computing capability. Alternatively, dynamic allocation mode could be used, which can achieve good performance in load balancing, but overhead may be increased, especially in this cluster mode, since the communication overhead among MPI nodes is much more than that in each node. As a result, in clusters, load balancing is not usually the main concern. Sometimes we can get better performance by static task allocation, although the computing capability may not be sufficiently used.

HPC Applications Based on MIC

11

By now we have learned the programming foundation on the MIC by having detailed development technology, processes, and optimization in the previous chapters. In this chapter, we explain the programming theory on MIC further with two actual cases in high-performance computing (HPC). First, we introduce a parallel algorithm for three-dimensional reconstruction of electron tomography based on the single-node CPU+MIC hybrid mode. Next we review a parallel algorithm of large eddy simulation based on the multi-node CPU+MIC hybrid mode, which will create a deeper comprehension of programming modes and optimization methods on the MIC.

The first case was developed by the Inspur Group and the Institute of Biophysics, Chinese Academy of Sciences. The sequential algorithm of this program is provided by Prof. Fei Sun's team in the Institute of Biophysics, Chinese Academy of Sciences. The MIC version has been designed and implemented by the Inspur Group in Beijing.

The second case was developed by the Inspur Group and Northwestern Polytechnical University, and the algorithm is provided by Chengwen Zhong's team. The MIC version of the program has also been designed and implemented by the Inspur Group.

Chapter Objectives.

- Learn the development process of HPC applications on the MIC
- Learn the development process of HPC applications on single-MIC, CPU+MIC hybrid computing on a single node, and CPU+MIC hybrid computing on multi-nodes
- Learn the implementation of CPU+MIC hybrid computing by MPI or OpenMP

E. Wang et al., *High-Performance Computing on the Intel® Xeon Phi™*,
DOI 10.1007/978-3-319-06486-4_11, © Springer International Publishing Switzerland 2014

11.1   Parallel Algorithms of Electron Tomography Three-Dimensional Reconstruction Based on Single-Node CPU+MIC Mode

11.1.1 Electron Tomography Three-Dimensional Reconstruction Technology and Introduction of SIRT Algorithms

In the field of life sciences, one of the most important method to study the non-uniform cells or the structure of macromolecules is electron tomography (ET). The simultaneous iterative reconstruction technique (SIRT) can bring the best performance in ET techniques.

11.1.1.1 Electron Tomography

Every scientific algorithm is based on some mathematical models or formulas. To better comprehend the SIRT program, it is necessary to learn the fundamentals of the mathematical model and the formulas behind its algorithm.

With electron tomography, we can reconstruct the body's three-dimensional (3D) inner structure by projection drawing from different angles. Simply put, the process involves placing the sample in the electron microscope and tuning the electron beam on the projection, drawing through the sample from the top to the bottom. One then obtains projection drawings from different angles by rotating the sample. The tilt range is around $\pm 60°$ to $\pm 70°$, and interval linear steps from $\pm 1°$ to $\pm 2°$ are usually adopted. First, we try to match these two-dimensional (2D) projection drawings from different angles, and then we reconstruct them using weighted back-projection or iterative reconstruction techniques to acquire the 3D structural drawing. As shown in Fig. 11.1, to match the resolution, the size of projection drawings is 1024*1024 points to 2048*2048 points.

In practice we reconstruct the body's structure with 2D cross-sections. The sample's fault section and its projection are shown in Fig. 11.2. Here, the lattice shows a section of the sample, and each little lattice stands for a pixel. Let x_j denote the value of the jth pixel, and N denote the total sum of the pixels. The line below denotes the projection of this section. P_i denotes one ray of the electron beam, whose value is the ray sum of the ith ray (the ray sum is the line integral of the rays). For the iterative reconstruction technique, the width of the electron beam is the same as one pixel. The area of the jth pixel crossed by ith ray (for example, A_{ij}, shown in Fig. 11.2) is called the weighted factor, and denotes the contribution of the jth pixel to the integral of the ith ray. Most of the A_{ij} are zero. The weighted factor can be solved by the area ratio of the lattice passed by the ray. Equation (11.1) shows the relationship between x and P.

$$\sum_{j=1}^{N} A_{ij} X_j = P_i, \quad i = 1, 2, 3, \ldots, M \tag{11.1}$$

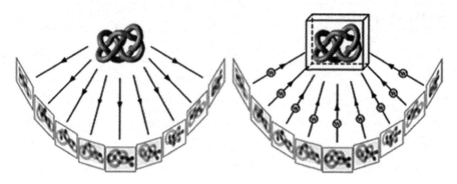

Fig. 11.1 Schematic diagram of electron tomography

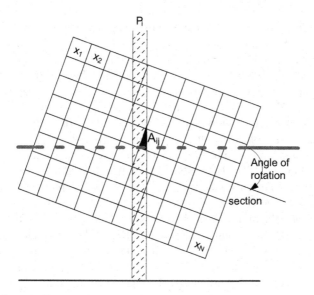

Fig. 11.2 Schematic diagram of the fault section and 2D projection

where N denotes the total sum of the pixels, and M denotes the total sum of the rays. If the values of M and N are small enough, we can use the direct method of solving linear equations. But, in practice, the value of N is always very large. To reconstruct a 1024*400 image, $N=409,600$, and M has the same magnitude as N. So, we can't apply the direct method for solving linear equations here. If the values of M and N are very large, we usually choose iterative methods. In this book, the SIRT algorithm is applied.

11.1.1.2 The Sequential SIRT

The value of each pixel is solved by accumulating the differences of the summation of all the rays which have passed through this pixel in SIRT [Eq. (11.2)]:

Fig. 11.3 Sample to be
reconstructed

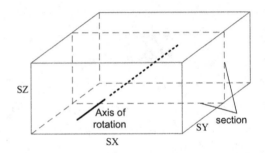

$$X^{(k)} = X^{(k-1)} + \lambda_k \sum_{i=1}^{M} \frac{P_i - A_i * X^{(k-1)}}{A_i * A_i} \qquad (11.2)$$

Here we assume that the length, width, and height of a sample are SX, SY, and SZ, respectively, and that the section is perpendicular to the SY direction. In other words, there are SY sections, and the area of each section is SX*SZ. Suppose the sum of the rotating angle is ANGLE_NUM, that is, there are ANGLE_NUM 2D projections, and the area of each projection is SX*SY (Fig. 11.3).

The sections are reconstructed one at a time. After all the sections are reconstructed in 2D, the 3D reconstruction image can be obtained. Reprojection and back-projection are the important procedures in the SIRT algorithm, and they consume a great deal of time in the reconstruction. In the reprojection procedure, the value of projection P_c can be calculated by reprojecting the initial value, $X^{(0)}$, or the updated value, $X^{(k-1)}$. In the back-projection procedure, the reconstructed image $X^{(k)}$ is reprojected by $X^{(0)}$, $X^{(k-1)}$, the measured projection P_t, and the calculated reprojected P_c. The main procedure of back-projection operates on the value X. We can obtain a high-quality reconstruction image by reprojection and back-projection. The procedure of sequential SIRT is shown in Fig. 11.4.

The sequential SIRT is shown in Algorithm 11.1, which contains three modules: initialization of back-projection, reprojection, and back-projection, which are shown in Algorithms 11.2, 11.3, and 11.4, respectively.

```
[pseudo code]
Begin
1     for: y=0:SY-1 //y denotes each section.
2         Read the projection image of yᵗʰ section.
3         Initialize the back-projection X⁽⁰⁾;
4         for: k=1: N //N denotes the iteration number
5             reproject; //Calculate Pc by X⁽ᵏ⁻¹⁾.
6             backproject; // Calculate X⁽ᵏ⁾ by Pc and Pt.
7         Return X⁽ᴺ⁾ of yᵗʰ section;
End
```

Algorithm 11.1   SIRT

```
[pseudo code]
Begin
1     for  n=0: ANGLE_NUM-1 //n denotes the number of the rotation angle.
2          for  z =0:  SZ-1 //z denotes the line number of the pixel in rejection image.
3               for  x = 0:  SX-1 //x denotes the column number of the pixel in rejection image.
4                    Confirm  the  ray(n, i),  which  is  across  pixel(z, x),  and  A(z, x)  by  the  rotation  angle  and
parameters;
5                         //(n, i) denotes the i^{th} ray in the n^{th} angle, and (n*SX+i)^{th} ray overall.
6                         // (z, x) denotes (z*SX+x)^{th} pixel in the reprojection image.
7                    X^{(0)}( z, x)  +=  A(z, x)*P_t(n, i);
End
```

Algorithm 11.2 Initialization of back-projection

```
[pseudo code]
Begin
1     for  n=0:ANGLE_NUM-1 // n denotes the number of the rotation angle.
2          for  i = 0:  SX-1 //i denotes the number of the pixel in the n^{th} angle in the projection image.
3               for zcur=0: SZ   //The loop step is cos( $\theta$ [n]) ,  $\theta$ [n] denotes the n^{th} rotation angle.
4                    Confirm the pixel (zcur, x), which is passed by ray(n, i), and A(zcur, x) by the rotation angle
and parameters;
5                    P_c(n, i) += A(zcur, x)*X^{(k-1)} (zcur, x);
End
```

Algorithm 11.3 Reprojection

```
[pseudo code]
Begin
1     for  n=0: ANGLE_NUM-1 // n denotes the number of the rotation angle.
2          for  z =0:  SZ-1 // z denotes the line number of the pixel in rejection image.
3               for  x = 0:  SX-1 // x denotes the column number of the pixel in rejection image.
4                    Confirm the ray(n, i), which is across pixel(z, x), and A(z, x) by the rotation angle and
parameters;
5                    X^{(k)} (z, x) += A(z, x)*(P_t(n, i)-P_c(n, i));
End
```

Algorithm 11.4 Back-projection

11.1.2 Analysis of the Sequential SIRT Program

Before porting and optimizing the sequential SIRT program, we should analyze the computational scale, the hotspot characteristics, the hotspot parallelism, vectorization, and memory needed on the MIC.

11.1.2.1 Computational Scale Analysis

Before designing the program flowchart, we must consider the scale of the application. The original program is sequential, and by the data scale and performance

Fig. 11.4 SIRT procedure

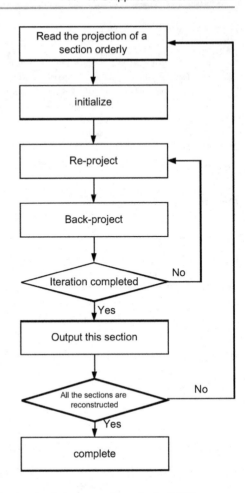

requirement, we can optimize by using one of these modes: single MIC on one node, multi-MIC on single node, or CPU+MIC hybrid computing. To help readers acquire a better understanding of developing procedures, we will introduce all of the three modes above. In this case, the size of the data needed for each section is several MB, and the time requirements are not too high. Therefore, the single node is sufficient, and there is no need to use a cluster.

11.1.2.2 Hotspot Characteristic Analysis

No matter which platform we want to work on, the hotspot characteristic analysis is most important. Then, with the help of some tools, we will analyze the sequential program to see if the hotspots are concentrated and find where any bottlenecks are located.

We proceed with the hotspot analysis on the SIRT algorithm using Intel VTune (Fig. 11.5). We can see that the initialization of back-projection, the reprojection, and the back-projection functions take 99% of the total time. So reprojection and

Call Stack	CPU Time▾	★	CPU Time:Total	Module	Function (Full)
▽ Total			100.0%		
▽ _start	0s		100.0%	recon_p...	_start
▽ main	0.010s		100.0%	recon_p...	main
reproject	286.920s		61.9%	recon_p...	reproject
backproject	155.920s		33.6%	recon_p...	backproject
back_project_init	20.640s		4.5%	recon_p...	back_project_init
mrc_add_slice	0.050s		0.0%	recon_p...	mrc_add_slice
mrc_update_head	0.020s		0.0%	recon_p...	mrc_update_head
slc_zero	0.010s		0.0%	recon_p...	slc_zero
▷ read_projection	0s		0.0%	recon_p...	read_projection
Selected 1 row(s):	0.050s		0.0%		

Fig. 11.5 Runtime of SIRT functions

back-projection are the hotspot functions in SIRT. In addition, although the initialization of back-projection takes only 5% of the time, to obtain improved performance on the MIC, we would also parallelize this function. This kind of the hotspot distribution is appropriate for parallelization because the hotspots are relatively concentrated, and if only these three functions were parallelized, the performance would be considered acceptable.

11.1.2.3 Hotspot Parallelism Analysis

Whether the hotspot functions can be parallelized depends on whether there is data dependency. In SIRT, there are all three loops (Algorithms 11.2, 11.3, 11.4) in the initialization of back-projection, reprojection, and back-projection functions. Because the structure of initialization of back-projection is the same as the back-projection function, we need only analyze the data characteristics of the reprojection and back-projection functions.

In reprojection function, to get the value P_c, there is no data dependency in the first and second outer loops of the three total. Thus they can readily be parallelized. The inner loop is the iterative accumulation process, which cannot be parallelized.

Similarly, in the back-projection function, to get the value $X^{(k+1)}$, there is no data dependency in the first and second outer loops of the three total. So they can be parallelized. The inner loop is again the iterative accumulation process, which cannot be parallelized.

11.1.2.4 Vectorization Analysis

By adding the -vec-report3 option when compiling, we can check whether the sequential codes were automatically parallelized by the compiler. By adding the -O3 option, we can see most of the array operations in the hotspot functions have

	Total Cores	Frequency	GDDR Size	GDDR5 Transfer Rate
MIC	61	1.1GHz	8GB	5.5GT/s

Fig. 11.6 MIC specification

ST image	Size (pixels)	Rotating angles	Thickness	Angle files
test4a	2048*2048 points	112	400	Get an angle value every 1° interval, begin with -60°

Fig. 11.7 Testing example

been parallelized automatically, which is appropriate to porting onto the MIC. The compiler outputs are shown below:

atom.c(698): (col.2) remark: LOOP WAS VECTORIZED.

atom.c(807): (col.2) remark: LOOP WAS VECTORIZED.

atom.c(911): (col.2) remark: LOOP WAS VECTORIZED.

11.1.2.5 Memory Analysis on MIC

Memory Size

In this case, the data size offloaded to the MIC is several MB, so that even with the memory occupancy of the system itself, the memory on the MIC is still enough. But if there is large memory requirement, and the running time is intolerable, we have to consider whether the data can be divided or whether we should try it on a cluster.

Memory Bandwidth

This is a computation-intensive example instead of a data-intensive one. So the memory bandwidth is not the bottleneck. Note: According to the analysis of the hotspot character, parallelism, and vectorization, we conclude that the SIRT algorithm is appropriate for parallelization. We expect to obtain good performance. Next, we will develop this parallel program by using OpenMP.

11.1.2.6 Experimental Platform

Hardware Environment

Server platform: Inspur NF5280M3 server

Specification: Dual Intel E5-2680 2.70GHz CPUs with 8 cores, 128GB memory DDR3, 2 *PCI-E x16 slots

The MIC specification is shown in Fig. 11.6:

Software Environment

OS: Red Hat Enterprise Linux 6.1 64bit

MIC Compiler: icc

The testing example is shown in Fig. 11.7:

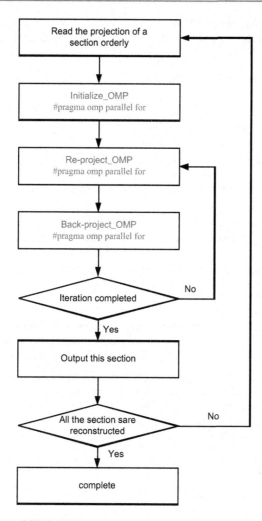

Fig. 11.8 Flowchart of SIRT_OMP

11.1.3 Development of a Parallel SIRT Program Based on OpenMP

We have found the hotspot functions in the SIRT program. According to the process of developing HPC applications on the MIC, we should rewrite the sequential program to an OpenMP version first, because OpenMP hyper-threading is supported in MIC. Then, we will develop and implement this OpenMP program in the following section.

OpenMP hyper-threading is based on shared memory. In this case, the OpenMP version is named SIRT_OMP for short. We implement the same functionality as SIRT_CPU using OpenMP, which can run on multi-core CPUs in the form of hyper-threading, and it will achieve a better performance. The solution is shown below in Fig. 11.8.

11.1.3.1 SIRT_OMP Algorithm

There are three modules in SIRT_OMP, which are the same as for SIRT_CPU:

- Initialization of the BackProject_Init
- ReProject
- BackProject

1. Initialization of the BackProject_Init

   This module initializes the cross-section X:

   Initialize the cross-section X with $X^{(0)}$ by back-projection. According to parallelism theory, we will only parallelize the outer loop. This means that "#pragma omp parallel for private(...) num_threads(THREAD_NUM)" will be added outside the outer loop, and THREAD_NUM threads will be utilized to implement parallelization. The inner loop will be accelerated automatically by vectorization. This process is shown in Algorithm 11.5.

	[pseudo code]
	Begin
1	#pragma omp parallel for private(n, z, x, ...) num_threads(THREAD_NUM)
2	for z = 0:SZ-1; //z is the line number of pixel in reconstructed image
3	for n = 0:ANGLE_NUM-1;// n is the order number of rotating angle
4	for x = 0:SX-1;//x is the column number of pixel in reconstructed image
5	Confirm A(z,x) and the ray (n,i), which is passed by pixel(z,x) according to the rotating angle and geometry parameter of the sample; //(n,i) denotes the ith ray of the nth rotating angle, which means it's the n*SX+Ith ray
6	//(z,x) denotes the z*SX+xth pixel in reconstructed image
7	Calculate $X^{(0)}(z,x)$ by the value of A(z,x) and $P_0(n,t)$;
	End

Algorithm 11.5 BackProject_Init_OMP

2. ReProject

   Calculate P_c by reprojection according to $X^{(k-1)}$. We just parallelize the outer loop (for(n=0: ANGLE_NUM-1)), and the inner loop is accelerated automatically through vectorization, shown in Algorithm 11.6.

> [pseudo code]
>
> Begin
>
> 1   #pragma omp parallel for private(n, I, zcur, ...), num_threads(THREAD_NUM)
>
> 2   for n = 0: ANGLE_NUM-1 //n denotes the order number of the rotating angle
>
> 3   for i = 0:SX-1 //i denotes the pixel order number of projection image in the ith angle
>
> 4   for zcur = 0:SZ // the step of loop is cos(θ[n]), θ[n] denotes the nth angle selected
>
> 5   Confirm the pixel (zcur, x) and A(zcur, x) passed by ray(n,i) according to the rotating angle and parameters
>
> 6 $$P_c(n,t)+ = A(zcur,x) * X(k-1)(zcur,x)$$
>
> End

Algorithm 11.6   ReProject_OMP

3. BackProject Module

  Calculate $X^{(k)}$ by $X^{(k-1)}$ and P_c according to the back-project algorithm. To reduce the overhead of the threads, we have adjusted the order of the loops, and only parallelize "for (z=0:SZ-1)". The algorithm is shown below (Algorithm 11.7).

> [pseudo code]
>
> Begin
>
> 1   #pragma omp parallel for private(ang_index, I, proindex_z_x, proindex0, proindex1,dx), num_threads(THREAD_NUM)
>
> 2   for z = 0:SZ-1 // z is the line number of pixel in reconstructed image
>
> 3   for n = 0:ANGLE_NUM-1;// n is the order number of rotating angle
>
> 4   for x = 0:SX-1;//x is the column number of pixel in reconstructed image
>
> 5   Confirm the rotating angle, the ray(n, i) passed by pixel (z, x) and A(zcur, x)
>
> 6 $$X(k)(z,x)+ = A(z,x) * P_t(n,i) - P_c(n,i);$$
>
> End

Algorithm 11.7   BackProject

11.1.3.2   Experimental Results

The performance results are shown in Fig. 11.9. For the sequential program, the performance is doubled after vectorization, while the OpenMP version has reached 22.55 times improvement. P_vec denotes the vectorized version.

11.1.4   Development of Parallel SIRT Programs Based on the MIC

By now we have finished the initial multi-thread version using OpenMP, and the performance increased by a large scale. But the number of CPU cores is no more

	Time (s)	Speedup
SIRT	3068	1x
SIRT_vec	1434	2.14x
SIRT_OMP	136	22.55x

Fig. 11.9 Performance of test4a OpenMP

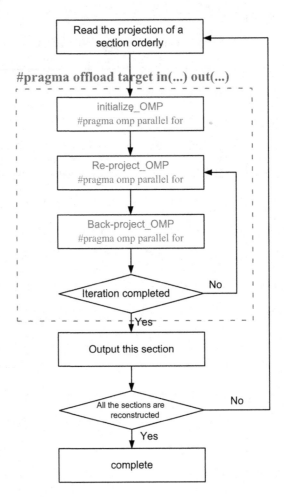

Fig. 11.10 Flowchart of SIRT_MIC

than 10–20, and the degree of parallelism is not good enough. So, we make use of
the MIC, which contains more cores to improve the performance of this case. The
MIC version of the program is named SIRT_MIC for short, and the flowchart is
shown in Fig. 11.10. To reduce the I/O operations between the CPU and the MIC,
we put the offload in front of the loop, and all the loops run on the MIC.

11.1.4.1 SIRT_MIC Algorithm

The parallelism methods of initialization, re-projection and back-projection are the same as the OpenMP version. In the initialization and back-projection, data is accessed by rows, so it can be vectorized automatically. While the data in re-projection is accessed by columns, then we must adjust the orders of the two inner loops to achieve vectorization. To vectorize the sequential and OpenMP programs, we add the forcing vectorization words "#pragma ivdep" in front of the inner loop.

11.1.4.2 Optimization Methods

1. Reduce the MIC offload

   As shown in Algorithm 11.1, in the initial stage, we call the offload statement in the initialization, reprojection, and back-projection functions. Then the number of times of data transfer between the CPU and the MIC is 1+2*N, while most of the data need not be transferred so frequently. In this way we can put the offload statement in front of the three functions, shown in Algorithm 11.8. In this way, the number of offload times is reduced from 1+2*N to 1, and the number of data transfer instances between the CPU and the MIC is bound to decrease.

[pseudo code]	
Begin	
1	for: y =0:SY-1 //y denotes each section
2	Read the value of projection image in the yth section
3	#pragma offload target in(...) out(...)
4	{
5	Initialization $X^{(0)}$;
6	for: k=1:N //N denotes the total number of the iterations
7	ReProject; //Calculate P_c by re-projection according to $X^{(k-1)}$
8	BackProject; // Calculate $X^{(k)}$ by back-projection according to P_c, P_t and $X^{(k-1)}$.
9	}
10	
11	Return $X^{(k)}$ after the yth section is reconstructed
End	

Algorithm 11.8   SIRT_MIC

2. Automatic Vectorization

   In the process of initialization, reprojection and back-projection functions, we just parallelize the outer loop in order to optimize the automatic vectorization. If we unroll the second layer of loops, only one element of the array is calculated by one thread; thus vectorization cannot be achieved. Hence we need to employ coarse-grained parallelism, and we only need to parallelize the outer loop. In the original program, after the outer loop is parallelized, the data in every thread in the initialization and back-projection functions is accessed by columns, which do not support automatic vectorization. So now we adjust the orders of the loops, as shown in Algorithms 11.9 and 11.10. Algorithm 11.9 is the original program, and Algorithm 11.10 is the new version after adjustment has been made. The new algorithm can proceed with the automatic vectorization.

[pseudo code]

Begin

1   for n=0:ANGLE_NUM-1; //n denotes the order number of the rotating angle

2   for i = 0:SX-1 //i denotes the pixel order number of projection image in the ith angle

3   for zcur = 0:SZ // the step of loop is cos(θ [n]), θ [n] denotes the nth angles selected

4   Confirm the pixel $X^{(k-1)}$ (zcur, x) and A(zcur, x) passed by ray(n,i) according to the rotating angle and parameters

5   Calculate $P_c(n,t)$ by $X^{(k-1)}$ (zcur, x) and A(zcur, x)

End

Algorithm 11.9 Re-Project

[pseudo code]

Begin

1   #pragma omp parallel for private(...), num_threads(THREAD_NUM)

2   for n = 0: ANGLE_NUM-1 //n denotes the order number of the rotating angle

3   for zcur = 0:SZ // the step of loop is cos(θ [n]), θ [n] denotes the nth angles selected

4   #pragma ivdep

5   for i = 0:SX-1;

6   Confirm the pixel $X^{(k-1)}$ (zcur, x) and A(zcur, x) passed by ray(n,i) according to the rotating angle and parameters

7   Calculate $P_c(n,t)$ by $X^{(k-1)}$ (zcur, x) and A(zcur, x)

End

Algorithm 11.10 Re-Project_vec

11.1.4.3 Experimental Results

With the same experimental platforms and data, the performance of test4a is shown in Fig. 11.11. We can see that the performance on the MIC is 29.79 times faster than that produced by the serial or sequential programs.

	Time (s)	Speedup
SIRT	3068	1x
SIRT_vec	1434	2.14x
SIRT_mic	103	29.79x

Fig. 11.11 Performance of test4a on MIC

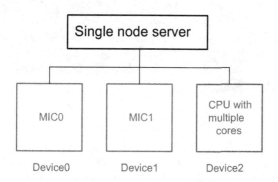

Fig. 11.12 Architecture of CPU+MIC system

11.1.5 Design of the Heterogeneous and Hybrid Architecture of CPU+MIC Mode Based on Single Nodes and Multiple Cards

Now we have completed porting and optimizing of SIRT based on a single MIC, we see that the performance has really improved. Although this version can be achieved quickly and easily, the CPU is left idle. So, if some of the workload can be allocated to the CPU, SIRT can be more efficient. Here we introduce CPU+MIC collaborative computing, which applies to larger-scale data. We have named this version SIRT_CPU_MIC for short.

In each single node, we adopt the desktop server with dual multi-core CPUs and two MIC cards, with the architecture shown in Figs. 11.12 and 11.13. In multi-MIC and CPU+MIC collaborative computing, we consider the dual CPUs as one device, and each MIC as a device. That is, there are three devices in every node, and each device is controlled by an OpenMP thread.

11.1.5.1 Task Division of Multi-MICs in Single-Node and CPU+MIC Collaborative Computing

In CPU+MIC collaborative computing, considering the character of the task, data, CPU and MIC, the task division is the most important. If there are M MICs in a single node and the dual CPUs are considered as devices, then we have $M+1$ devices in our case. We divide the tasks for a well-balanced equilibrium, in which each device obtains the input data from every section automatically, and

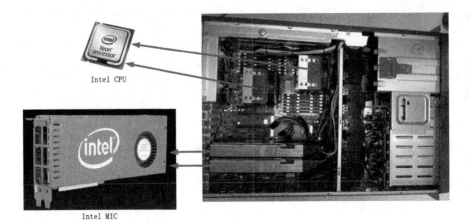

Intel CPU

Intel MIC

Fig. 11.13 Architecture of the server

each device obtains the next input data from the next section after the current computation, until the whole device finishes calculating data from all sections.

11.1.5.2 Structure of Multi-MIC in Single-Node and CPU+MIC Collaborative Computing

We show in Fig. 11.14 the framework of CPU+MIC collaborative computing after task division.

In a single node, there are $M+1$ devices. If we adopt the fork-join mode by OpenMP as our framework in the node, there is only one main thread when the program starts, and the additional threads will be generated when the parallelism is required. This means that we start $M+1$ OpenMP threads, and control the dynamic I/O of each thread by a while(1) loop. The threads $0 \sim M-1$ control the whole MIC device, and the thread M controls the CPU.

In this case, in the CPU+2MIC platform, the main thread controls the dynamic distribution of the input data. Threads 0, 1, and 2 controls MIC0, MIC1, and CPU, respectively (Fig. 11.14).

In the SIRT algorithm, there is enough memory on the MIC to store the section data, so it is not necessary to partition them. As shown in Fig. 11.14, after each device reconstructs data from a section, the reconstruction image is generated. When reading the current section, the lock operation is necessary to ensure the independence of every section in order to avoid conflicts in read and write. As a result, there is no dependence among these devices.

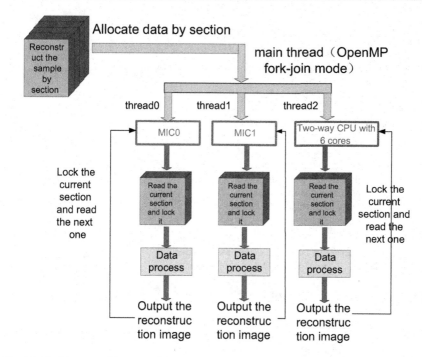

Fig. 11.14   Diagram of CPU+MIC collaborative computing in a single node

The pseudo-code of the CPU+MIC collaborative computing framework in single node is shown as below:

```
[code snippet]
//Begin
1    //SY denotes the number of sections.
2    int DEVICE_NUM=N  //N denotes the number of all the CPU and MIC.
3    int Slice_private=0;   //Define a private section variable for each thread.
4    int Slice=0;       //Define a shared variable to control the reading and writing of the section.
5    omp_lock_t lck;
6
7    #pragma omp parallel for private (Slice_private) ,num_threads (DEVICE_NUM)
8    for (thread=0; thread<DEVICE_NUM; thread++)
9    {
10   while(1) //The endless loop controls the N threads to working all the time, and locks them when one thread is
     reading a section to avoid a conflict.
11   {
12     if( Slice>SY )
13     {
14             break;
15     }
16     omp_set_lock (&lck);
17     {
18             Slice_private=slice;
19             Slice++; //Shared varible++
20     }
```

```
21     omp_destory_lock(&lck);
22
23     omp_set_lock (&lck);
24     Read(Slice_private);//Read the project image of slice_private^th section.
25     omp_destory_lock(&lck);
26     /* If the thread number is less or equal to DEVICE_NUM-1, it will be running on MIC, or it'll run on
CPU.*/
27     if( thread <DEVICE_NUM  ) //mic 0,mic 1 compute
28     {
29              #pragma offload target(mic:0)\
30          in(SX, SZ, ANGLE_NUM, N, …) \
31          in (P_t: length(ANGLE_NUM * SX)) \
32          in (thita: length(ANGLE_NUM)) \
33          in( P_c: length(ANGLE_NUM * SX)) \
34          inout (X: length(SZ*SX))

35          {
36              BackProject_Init();  //Initialize the back-projection X^{(0)}
37                   for: k=1: N // N denotes the number of iterations.
38                        ReProject();   //re-projection
39                        BackProject(); //back-projection
40          }
41          Return X(N) after the slice_private^th section re-projected;
42     }
43     else if(thread==DEVICE_NUM-1) //cpu compute
44     {
45              BackProject_Init();  //Initialize back-projection by X(0)
46              for: k=1: N //N denotes the number of iterations.
47                   ReProject();   //re-projection
48              BackProject();  //back-projection
49              Return X(N) after the slice_private^th section re-projected;
50     }
51     else   //other mic compute
52     {
53              #pragma offload target (mic:1)\
54              in(SX, SZ, ANGLE_NUM, N, …) \
55              in (P_t: length(ANGLE_NUM * SX)) \
56              in (thita: length(ANGLE_NUM)) \
57              in (P_c: length(ANGLE_NUM * SX)) \
58              inout (X: length(SZ*SX))
59              {
60                  BackProject_Init();   // Initialize back-projection by X(0)
61                       for: k=1: N // N denotes the number of iterations.
62                       ReProject();     // re-projection
63                       BackProject();  // back-projection
64              }
65              Return X(N) after the slice_private^th section re-projected;
66     }
67  }  // end while
//End
```

This framework can be used in multi-MIC computation and in CPU+MIC collaborative computing. However, we must adjust the variable DEVICE_NUM, which is the number of devices, or trim the if ... else ... codes, so the framework can run without a hitch.

When the data input codes are all before the if ... else ... structure:

1. If we continue with the collaborative computing, the variable DEVICE_NUM is considered to include the number of CPU and MIC; in the meantime, there are DEVICE_NUM OpenMP threads to control these devices. As shown in the above codes, there is no need to modify the if ... else ... structure.
2. If we continue with the multi-MIC computation, we consider DEVICE_NUM to include only the number of MICs. We should modify "if (thread<DEVICE_NUM)" to "if(thread<=DEVICE_NUM)"; then the CPU-side computation will not run, while the multi-MIC can still run.

11.1.5.3 Load Balancing of Multi-MIC in Single-Node and CPU+MIC Collaborative Computing

Because the computation capability of CPU cores is different from that of MIC cores, the load balancing of CPU+MIC collaborative computing is crucial. The task-level and algorithm-level parallelisms in SIRT should be compared to find feasible load balancing solutions.

1. Compare task-level and algorithm-level parallelisms in SIRT
   (a) Task-level parallelism in SIRT
      As shown in Fig. 11.14, the 3D projection data is shown with a cross-section. Because the format and size of each section are all the same, the running time is nearly the same. So whether the computation goes on either the CPU core or the MIC card, it is feasible to proceed by each section. The computation capability of each core in the CPU or in the MIC is the same, so if the data for each section is allocated to the cores in the CPU or in the MIC, we can achieve good parallelism performance.
   (b) Algorithm-level parallelism in SIRT
      Divide the data for each section, and then allocate to different CPU cores or MIC cards to proceed with parallelization of the section. According to the analysis of the SIRT algorithm, there are interactive data to deal with. This means that there is a large amount of I/O transfer, which seriously impacts the performance.
2. A load balancing solution for CPU+MIC collaborative computing
      In multi-MIC and CPU+MIC collaborative computing, DEVICE_NUM OpenMP threads are started by the while(1) loop, and each thread carries out the reading, writing and distribution of the input data dynamically. In SIRT, the memory on the MIC is sufficient to store the section data, thus there is no need to partition them. The reconstruction image is generated after each device reconstructs data for a section. When reading and writing data for the current section, the lock operation is necessary to ensure the independence of every section and to solve any read and write conflicts. As a result there is no dependence among these devices.

	Time(s)	Speedup
SIRT	3068	1x
SIRT_vec	1434	2.14x
SIRT_CPU_MIC	41	74.83x

Fig. 11.15 Performance of test4a (MIC version)

11.1.5.4 Experimental Results

The performance of test4a using the same platform and testing data is shown in Fig. 11.15. Compared to the serial version of the code, the MIC version is 74 times faster.

11.2 Parallel Algorithms of Large Eddy Simulation Based on the Multi-node CPU+MIC Mode

11.2.1 Large Eddy Simulation Based on the Lattice Boltzmann Method

11.2.1.1 Lattice Boltzmann Method

The Lattice Boltzmann method (LBM) has developed into a widely used numerical algorithm over the past 20 years. It is based on both microscopic molecular dynamics and on macroscopic continuum mechanics hypotheses. This method is different from traditional fluid simulation methods, since it is derived from molecular motion theory. By tracking the trace of the molecular distribution function, the macroscopic uniform characteristic of the fluid is obtained after solving for the distribution function. Molecular motion theory, which is so characteristic of LBM, makes it more effective to simulate some complicated fluid, such as porous media, suspension flow, multiphase flow, multi-component flow, and so on. Good parallelism exists in LBM itself, and the boundary communications can be seamlessly implemented.

The LBM method of computational fluid dynamics is different from the traditional numerical algorithm and is a discrete way to solve the Boltzmann equation. The whole process is time-dependent and has a good spatial dependence, which is appropriate for parallelism.

There are two steps to solve the Lattice Boltzmann equation:

1. Collision term:

$$f_i^*(x,t) = f_i(x,t) - \frac{1}{\tau}\left[f_i(x,t) - f_i^{(eq)}(x,t)\right] \tag{11.3}$$

2. Migration term:

$$f_i(x, t + \Delta t) = f_i^*(x - c_i \Delta t, t) \tag{11.4}$$

Fig. 11.16 D2Q9 lattice model

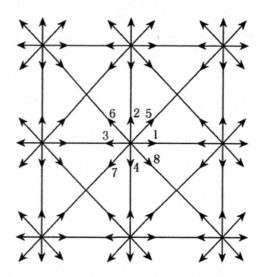

where i denotes the different directions of a particle's discrete velocity, x denotes the location of the discrete nodes, $f_i(x,t)$ denotes the particle distribution function of discrete velocity direction I, Δt is the time step, c_i is the particle velocity in discrete direction, $f_i^*(x,t)$ is the equilibrium distribution function after the collision. Here τ denotes the delay time; the equation for τ is given by:

$$\tau = \frac{3U_0L}{Rec^2\Delta t} + 0.5 \tag{11.5}$$

where U_0 is the characteristic velocity, L is the characteristic length, and Re is the Reynolds number, which is the ratio of inertial force and viscous force.

In the D2Q9 model of LBM shown in Fig. 11.16, the equilibrium distribution function is:

$$f_i^{(eq)}(x,t) = w_i\rho\left[1 + \frac{3c_{i,\alpha}u_\alpha}{c^2} + 9\frac{(c_{i,\alpha}u_\alpha)^2}{2c^4} - \frac{3u^2}{2c^2}\right] \tag{11.6}$$

where $w0/=4/9$, $w1=...=w4=1/9$, $w5=...=w8=1/36$; there are nine equilibrium distribution functions in different directions. The macroscopic variables on lattice, like density, pressure, and velocity, can be obtained from these distribution functions, and the equations are:

$$\rho = \sum_i f_i, \quad p = \frac{\rho}{3}c^2, \quad u = \frac{1}{\rho}\sum_i f_ic_i \tag{11.7}$$

where $c_s = \frac{c}{\sqrt{3}}$ is the sound velocity.

Fig. 11.17 LBM-LES submodule

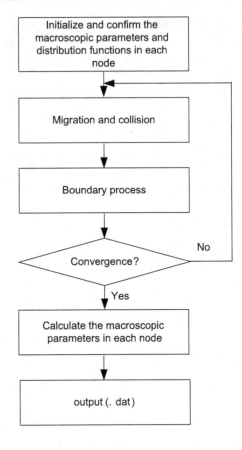

The boundary condition plays an important role in numerical precision and stability in LBM applications. In this case, we adopt the nonequilibrium extrapolation boundary conditions method to solve the solid-wall boundary conditions.

The basic idea is to divide the unknown distribution functions into equilibrium state and nonequilibrium state, and then the nonequilibrium part can be obtained by first-order precision extrapolation. This can be denoted by:

$$f_i^0 - f_i^{0.(eq)} = f_i^1 - f_i^{1.(eq)} \tag{11.8}$$

where f_i^0, f_i^1 denote the distribution functions on the lattice in an actual boundary and outside an adjacent boundary, and $f_i^{0.(eq)}$, $f_i^{1.(eq)}$ are the corresponding equilibrium distributions.

11.2.1.2 Large Eddy Simulation

Large eddy simulation (LES for short) has been developed as a numerical method in fluid dynamics in the past 10 years; it differs from traditional methods such as direct numerical simulation (DNS) and Reynolds-averaged Navier-Stokes (RANS) methods. In LES, all the movement above the turbulence level is simulated accurately to model the time-dependent evolution of large-scale dynamic effects in non-equilibrium situations, and coherent structure, which the RANS method cannot resolve. In addition, it overcomes the disadvantage of large computation overhead in modeling turbulence. Today LES has become one of the most influential numerical methods in turbulence simulation.

The LBM can be used for solving LES. Then we can get the LBM of large eddy simulation (LBM-LES) by combining the two together, which is widely popularized in fluid dynamics.

LBM-LES contains five modules (Fig. 11.17).

1. Initializing module

   In LES algorithms, some variables and arrays are initialized in this module, like the size of grid(nx*ny), Reynolds number(Re), initial velocity(U0), velocity (ux, uy, the sizes of which are both nx*ny), double-buffering distribution function(fr0, fe0, fn0, fw0, fs0, fne0, fnw0, fsw0 , fse0, and fr1, fe1, fn1, fw1, fs1, fne1, fnw1, fsw1, fse1; the sizes of which are nx*ny, r, e, n,... in 9 directions), density(rho, the size of which is nx*ny), stream(stream, the size of which is nx*ny), which is used for storing the output.

2. Migration and collision module

   We implement the migration and collision in one function (LBCollProp). The equations of migration and collision are shown in Eqs. (11.3) and (11.4), and the pseudocode of migration-collision is:

```
[code snippet]
1     void LBCollProp(parameter...) //in: f*0, out: f*1. * denotes r, e, n, w, s, ne, nw, sw, se, respectively.
2     {
3          for(i=1;i<ny-1;i++)
4               for(j=1;j<nx-1;j++)
5               {
6                    /*migration*/
7                    fr = fr0(i)(j);   // f*(i)(j) denotes f*[i][j].
8                    fe = fe0(i)(j-1);
9                    fn = fn0(i-1)(j);
10                   fw = fw0(i)(j+1);
11                   fs = fs0(i+1)(j);
12                   fne = fne0(i-1)(j-1);
13                   fnw = fnw0(i-1)(j+1);
14                   fsw = fsw0(i+1)(j+1);
15                   fse = fse0(i+1)(j-1);
16                   /*collision*/
17                   f*1(i)(j) = fun(fr, fe, fn, fw, fs, fne, fnw, fsw, fse, ...);
18               }
19    }
```

3. Boundary module

In this module we have adopted a non-equilibrium extrapolation method, which is divided into left–right module, upper module, and lower module. The pseudo-code is shown below:

```
[code snippet]
1    /*" Non-equilibrium extrapolation method " solve the boundary lattice. The left-right, upper , and lower
     boundary will be solved, respectively.*/
2    // left-right boundary module
3    void LBBC_LR(parameter...) //in: f*0, out: f*1. * denotes r, e, n, w, s, ne, nw, sw, se respectively.
4    {
5        // left boundary
6        j = 0;
7        for(i=0;i<ny;i++)
8        {
9            f* = f*(i)(j+1);
10           f*1(i)(j) = fun(fr, fe, fn, fw, fs, fne, fnw, fsw, fse, ...);
11       }
12       // right boundary
13       j = nx-1;
14       for(i=0;i<ny;i++)
15       {
16           f*(i)(j) = f*(i)(j-1);
17           f*1(i)(j) = fun(fr, fe, fn, fw, fs, fne, fnw, fsw, fse, ...);
18       }
19   }
20
21   //upper boundary
22   void LBBC_UP(parameter...) //in: f*0, out: f*1. * denotes r, e, n, w, s, ne, nw, sw, se respectively.
23   {
24       j = ny-1;
25       for(j=0; j<ny; j++)
26       {
27           f* = f*(i-1)(j);
28           f*1(i)(j) = fun(fr, fe, fn, fw, fs, fne, fnw, fsw, fse, ...);
29       }
30   }
31
32   //lower boundary
33   void LBBC_ DOWN (parameter...) //in: f*0, out: f*1. * denotes r, e, n, w, s, ne, nw, sw, se, respectively.
34   {
35       j = 0;
36       for(j=0; j<ny; j++)
37       {
38           f* = f*(i+1)(j);
39           f*1(i)(j) = fun(fr, fe, fn, fw, fs, fne, fnw, fsw, fse, ...);
40       }
41   }
```

THE migration-collision module and the boundary module are iterative processes, which may run for many iterations.

4. Macroscopic parameter computation module

After the migration-collision module and boundary module have completed, this module is used to compute the macroscopic parameters in all the grids

5. Output module

LES algorithm outputs the results in dat format.

The pseudo-code of the LES algorithm structure is shown below:

```
[pseudo-code]
1    void main()
2    {
3          init(); //Initialization module
4          for(i=0; i<ITR; i++)
5          {
6                LBCollProp(); //migration and collision
7                LBBC_LR(); //left-right boundary
8                LBBC_UP(); //upper boundary
9                LBBC_DOWN(); //lower boundary
10         }
11         computeMacroparam (); //macroscopic parameters computation module
12         output(); //results
13   }
```

11.2.2 Analysis of Large Eddy Simulation Sequential (Serial) Program

11.2.2.1 Computational Scale Analysis

In the large eddy simulation (LES) algorithm, the sizes of the grids are all nx*ny, the time complexity is O(nx*ny), and the space complexity is O(nx*ny). The computational complexity rises quickly with increasing grid size. In current fluid dynamics research, the grid size has grown larger and larger, from millions to billions, and even more. A single computer cannot meet the computational demands of extremely high resolution in both time and space. Therefore we now need to use clusters. In this subsection we introduce the solution of the LES model over large grid systems using CPU+MIC collaborative computing.

11.2.2.2 Hotspots Characteristic Analysis

The key to program optimization is the hotspots. Before optimizing the LES, we should test its hotspots. Through VTune, we find the sequential LES hotspots (Fig. 11.18). We can see that 99% of the time is spent on the solution of migrations-collisions; this is our hotspot. In addition, the computation of the lower boundary, the left-right boundary, and the upper boundary together are less

Call Stack	CPU Time▾	CPU Time:Total	Module	Function (Full)
▽Total		100.0%		
▽_start	0s	100.0%	les	_start
▽main	0s	100.0%	les	main
▽benchMark	0s	100.0%	les	benchMark(voi ...
▽evolution	0s	100.0%	les	evolution(int)
▷launch_LBCollProp	0s	99.1%	les	launch_LBColl...
▷launch_LBBC_DOWN	0s	0.1%	les	launch_LBBC_ ...
▷launch_LBBC_LR	0s	0.5%	les	launch_LBBC_ ...
▷launch_LBBC_UP	0s	0.2%	les	launch_LBBC ...

Hotspots - Hotspots
Analysis Target | Analysis Type | Collection Log | Summary | Bottom-up | Top-down Tree

Fig. 11.18 The hotspots of the LES sequential (serial) program found by VTune

than 1%. There are few I/O operations, and only some macroscopic parameter inputs showing up in the VTune results.

11.2.2.3 Hotspots Parallelism Analysis

We found the LES sequential program hotspots in the last section; now we analyze the parallelism of the hotspot functions. Migration-collision is the hotspot function in the LES sequential program. The value of the equilibrium distribution function of each node in the 2D grid is calculated by this function, and there is no dependence in the solution, which corresponds to the two levels of loops in the LBCollProp function. Hence, the two levels of loops can be readily parallelized. According to the parallelization principle, we can parallelize the outer loop, and vectorize the inner loop for optimization.

In addition, the boundary function is used to compute the equilibrium distribution function value of every boundary node, and there is also no data dependence. Hence it can also be parallelized. Although it is not a hotspot function, the parallelism of this function should also be analyzed in order to run these functions uniformly on the MIC and to reduce the communication between the CPU and the MIC. Later, we will introduce in greater detail the advantage of putting the boundary function on the MIC.

11.2.2.4 Vectorization Analysis

In the chapter on vectorization optimization we introduced the compiler options: we can check the vectorization degree of sequential LES by adding the "-vec-report3" before compiling. As shown in Fig. 11.19, we can see that there is a large number of loops that have not been completely vectorized in the serial version of LES. Thus we should add some directives of automatic vectorization or modify the loops to increase the degree of vectorization.

11.2.2.5 MIC Memory Analysis

Our MIC memory analysis focuses on two aspects: memory size and memory bandwidth.

```
les_kernel.cpp(47): (col. 3) remark: loop was not vectorized: existence of vector dependence.
les_kernel.cpp(46): (col. 2) remark: loop was not vectorized: not inner loop.
les_kernel.cpp(74): (col. 3) remark: loop was not vectorized: existence of vector dependence.
les_kernel.cpp(73): (col. 2) remark: loop was not vectorized: not inner loop.
les_kernel.cpp(106): (col. 3) remark: loop was not vectorized: existence of vector dependence.
les_kernel.cpp(106): (col. 3) remark: loop skipped: multiversioned.
les_kernel.cpp(102): (col. 2) remark: loop was not vectorized: not inner loop.
les_kernel.cpp(431): (col. 2) remark: loop was not vectorized: existence of vector dependence.
les_kernel.cpp(431): (col. 2) remark: loop was not vectorized: not inner loop.
les_kernel.cpp(438): (col. 2) remark: loop was not vectorized: existence of vector dependence.
les_kernel.cpp(438): (col. 2) remark: loop was not vectorized: not inner loop.
les_kernel.cpp(445): (col. 2) remark: loop was not vectorized: existence of vector dependence.
les_kernel.cpp(445): (col. 2) remark: loop skipped: multiversioned.
les_kernel.cpp(445): (col. 2) remark: loop was not vectorized: not inner loop.
les_kernel.cpp(161): (col. 3) remark: loop was not vectorized: existence of vector dependence.
les_kernel.cpp(160): (col. 2) remark: loop was not vectorized: not inner loop.
les_kernel.cpp(232): (col. 2) remark: loop was not vectorized: existence of vector dependence.
les_kernel.cpp(270): (col. 2) remark: loop was not vectorized: existence of vector dependence.
les_kernel.cpp(319): (col. 2) remark: loop was not vectorized: existence of vector dependence.
les_kernel.cpp(378): (col. 2) remark: loop was not vectorized: existence of vector dependence.
les_kernel.cpp(123): (col. 3) remark: loop was not vectorized: existence of vector dependence.
les_kernel.cpp(122): (col. 2) remark: loop was not vectorized: not inner loop.
```

Fig. 11.19 The degree of vectorization in the serial LES code

Memory Size

When solving a large number of grid points by LES, a great deal of memory is occupied. The amount of memory occupied is directly related to the size of the grid points. For example, the equilibrium distribution value in nine directions will be stored in each grid point, and requires double-buffering to store the input and output. For floating point operations, 1 billion grid points would require 72 GB memory. So LES requires large amounts of memory, and it is better to run in clusters in the case of large grid sizes.

Memory Bandwidth

There are many iterations in LES. During each of these iterations the memory is accessed frequently, and the computation is intensive. Thus LES is appropriate for the MIC.

11.2.3 Parallel Algorithm of Large Eddy Simulation Based on OpenMP

11.2.3.1 Multi-threading Parallelism by OpenMP

According to the programming process on MIC, before porting a program to MIC, the OpenMP version of the program is needed. In this subsection, we introduce the parallel algorithm of LES based on OpenMP.

From the analysis of the hotspots and parallelism of sequential LES, we need to implement the parallelization of the migration-collision function. This means that we should parallelize that function by OpenMP. In the meantime, we will parallelize the OpenMP version of the boundary function to optimize communication on the MIC.

There are two levels of loops in the migration-collision function, and there are data dependencies in both of them. According to the parallelism principle, to implement the coarse-grain parallelism, we will parallelize the outer loop and use directives (like #pragma ivdep) in the inner loop of the migration-collision function. The pseudo-code of the migration-collision function in OpenMP is shown below:

```
[pseudo-code]
1    void LBCollProp(parameter...) //in: f*0, out: f*1. * denotes r, e, n, w, s, ne, nw, sw, se respectively.
2    {
3    #pragma omp parallel for private(i, j, ...) num_threads(THREAD_NUM)
4        // THREAD_NUM denotes the number of started threads.
5        for(i=1;i<ny-1;i++)
6            #pragma ivdep //auto-vectorization directive
7            for(j=1;j<nx-1;j++)
8            {
9                /*migration*/
10                   ...
11               /*collision*/
12                ...
13            }
14    }
```

There is one level of loop in the boundary function, so we can only parallelize this loop. The pseudo-code of this parallelism by OpenMP is:

```
[pseudo-code]
1    //left-right boundary
2    void LBBC_LR(parameter...) //in: f*0, out: f*1. * denotes r, e, n, w, s, ne, nw, sw, se respectively.
3    {
4        // left boundary
5        j = 0;
6        #pragma omp parallel for private(i, ...) num_threads(THREAD_NUM)
7        for(i=0;i<ny;i++)
8        {
9            ...
10        }
11        // right boundary
12        j = nx-1;
13        #pragma omp parallel for private(i, ...) num_threads(THREAD_NUM)
14        for(i=0;i<ny;i++)
15        {
16            ...
17        }
18    }
19
```

```
20    //upper boundary
21    void LBBC_UP(parameter...) //in: f*0, out: f*1. * denotes r, e, n, w, s, ne, nw, sw, se respectively.
22    {
23          j = ny-1;
24          #pragma omp parallel for private(j, ...) num_threads(THREAD_NUM)
25          for(j=0; j<ny; j++)
26          {
27                ...
28          }
29    }
30
31    //lower boundary
32    void LBBC_ DOWN (parameter...) //in: f*0, out: f*1. * denotes r, e, n, w, s, ne, nw, sw, se respectively.
33    {
34          j = 0;
35          #pragma omp parallel for private(j, ...) num_threads(THREAD_NUM)
36          for(j=0; j<ny; j++)
37          {
38                ...
39          }
40    }
```

11.2.3.2 Experimental Results

Experimental environment

The hardware environment of testing the LES application is shown in Fig. 11.20. The performance of LES is counted by LUPS (Lattice Unit Per Second), and usually this is MLUPS (Million Lattice Unit Per Second). The equation is:

$$P = NX^*NY^*S/T \tag{11.9}$$

where NX, NY are the width and height of the grid cell respectively, S is the number of iterations of the fluid field, T is the computing time, and P is the update rate of the lattice.

Experimental results

The experimental results of LES on the multi-core CPU are found in Fig. 11.21, and the speedup is shown in Fig. 11.22. The speedup of the multi-threading LES on the 16-core CPU platform is more than 20. The reason for the superlinear speedup is that we have vectorized the LES of the OpenMP version, and the vectorization degree has been increased. The vectorization degree of OpenMP version is shown in Fig. 11.23.

Platform	Inspur NF5280M3
CPU	Intel Xeon CPU E5-2680 2.7GHz, two-way with 8 cores
Memory	DDR3 1333MHz 128GB
MIC	KNC, 61 cores, 1.1GHz, GDDR5 8GB memory 5.5GT/s
OS	Red Hat Enterprise Linux Server release 6.1, 64bit
Compiler	icc
Sample	Reynolds number: 10000; Iteration: 10000

Fig. 11.20 LES experimental environment

	MLUPS				
	512*512	1024*1024	2048*2048	4096*4096	8192*8192
Sequential CPU	12.49	12.09	9.11	8.97	8.5
OpenMP multithread	280.02	243.28	198.52	199.77	198.84

Fig. 11.21 Experimental results of parallel LES based on OpenMP

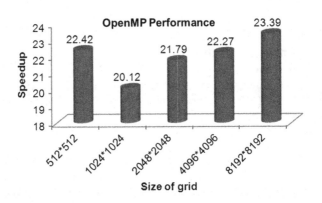

Fig. 11.22 Speedup of parallel LES based on OpenMP

11.2.4 Parallel Algorithm of Large Eddy Simulation Based on MIC

11.2.4.1 LES_MIC

We have implemented the multi-thread parallel LES based on OpenMP. Now we can port it over to the MIC. As shown in Fig. 11.24, the migration-collision module will be moved to MIC. The pseudo-code of parallel LES based on MIC is shown in the algorithm LES_MIC.

Algorithm LES_MIC

```
[pseudo-code]
1    void main()
2    {
3        init(); // Initialization
4        /*Allocate space on MIC, and transfer the input data without releasing the space*/
5    #pragma offload target(mic) \
6        in(fr0, fn0, ..., fse0: length(nx*ny) alloc_if(1) free_if(0)) \
7        nocopy(fr1, fn1, ..., fse1: length(nx*ny) alloc_if(1) free_if(0))
8        {}
9        for(i=0; i<ITR; i++)
10       {
11           /*kernel computation section, there's no data transfer in each iteration*/
12    #pragma offload target(mic) \
13           nocopy(fr0, fn0, ..., fse0) \
14           nocopy(fr1, fn1, ..., fse1)
15           {
16               LBCollProp(); //migration and collision
17               LBBC_LR(); //left-right boundary
18               LBBC_UP(); //upper boundary
19               LBBC_DOWN(); //lower boundary
20           }
21       }
22       /*Return results and release the space allocated on MIC.*/
23       if(ITR%2==0)
24       {
25    #pragma offload target(mic) \
26           out(fr0, fn0, ..., fse0: length(nx*ny) alloc_if(0) free_if(1)) \
27           nocopy(fr1, fn1, ..., fse1: length(nx*ny) alloc_if(0) free_if(1))
28           {}
29       }
30       else
31       {
32    #pragma offload target(mic) \
33           nocopy(fr0, fn0, ..., fse0: length(nx*ny) alloc_if(0) free_if(1)) \
34           out(fr1, fn1, ..., fse1: length(nx*ny) alloc_if(0) free_if(1))
35           {}
36       }
37       computeMacroparam (); //macroscopic parameters computation module
38       output(); //results
39   }
```

```
les_kernel.cpp(170): (col. 3) remark: LOOP WAS VECTORIZED.
les_kernel.cpp(242): (col. 2) remark: LOOP WAS VECTORIZED.
les_kernel.cpp(281): (col. 2) remark: LOOP WAS VECTORIZED.
les_kernel.cpp(331): (col. 2) remark: LOOP WAS VECTORIZED.
les_kernel.cpp(391): (col. 2) remark: LOOP WAS VECTORIZED.
```

Fig. 11.23 Degree of vectorization of parallel LES based on OpenMP

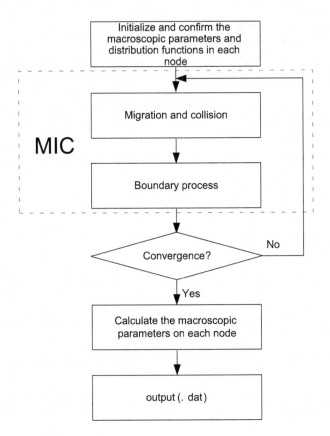

Fig. 11.24 LES_MIC framework

We have applied nocopy technology to reduce the data communication. Memory space on the MIC is allocated for equilibrium function f*0, f*1(* denotes r, e, n, w, s, ne, nw, sw, se) without release in lines 4–8; in the meantime, the data in f*0 is transferred from the CPU to the MIC. The computation part on the MIC is in lines 12–20. Because the data is on the MIC all the time, f*0 and f*1 use a nocopy method instead of advancing the I/O operations in each iteration. The computational result is transferred from the MIC back to the CPU, and the memory allocated on the MIC is released in lines 23–36. We can certainly move the offload words in

	MLUPS				
	512*512	1024*1024	2048*2048	4096*4096	8192*8192
Sequential CPU	12.49	12.09	9.11	8.97	8.5
OpenMP multithread	280.02	243.28	198.52	199.77	198.84
MIC multithread	27.49	87.16	203.74	309.56	351.7

Fig. 11.25 Experimental results of parallel LES based on OpenMP

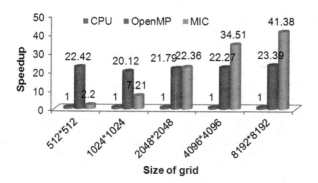

Fig. 11.26 Speedup of LES_MIC

line 12 in front of the iteration in line 9; hence we do not need the nocopy. But we still put the offload words in the iterations for the multi-node and multi-card computation.

11.2.4.2 Experimental Results on MIC

The performance on a KNC MIC is shown in Fig. 11.25. When the size of the grid is 8192*8192, the performance on the MIC is 351.7MLUPS. The speedup is shown in Fig. 11.26, and the highest speedup is 41.38.

11.2.5 Parallel Algorithm of Large Eddy Simulation Based on Multi-nodes and CPU+MIC Hybrid Platform

In the last section, we concluded that a single computer cannot meet the demands of computation both in time and space for LES, so we need to consider the CPU+MIC collaborative computing on multiple nodes. In this section, we introduce this algorithm and the development on the CPU+MIC.

11.2.5.1 Parallelism Design

There are three levels of parallelism in the parallel LES on the CPU+MIC. First, MPI is used for the data partition and message passing among nodes. Second, MPI is also used for the data partition and message passing in the current node for both

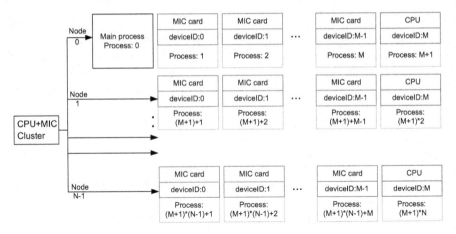

Fig. 11.27 Structure of CPU+MIC collaborative computing on multiple nodes

the CPU and multiple MICs. Finally, the innermost parallelism is multi-thread optimization based on OpenMP on the CPU or the MIC. The parallelism both on the node and among nodes is based on MPI, thus we can combine them.

The framework of parallel LES on CPU+MIC is shown in Fig. 11.27. We assume there are N nodes, and in each of them there are M MIC cards. The whole CPUs in each node are considered as one single device. We start N*(M+1)+1 processes, in which process 0 is the main process to allocate data and proceed the I/O operations. Additional N*(M+1) processes are used for computation, and each of them controls one CPU or MIC (Fig. 11.27).

The flowchart of CPU+MIC collaborative computing is shown in Fig. 11.28. The main process broadcasts data to N*(M+1) MPI processes, and then every process calls the CPU or the MIC to advance computation. In the case of the MIC, the process offloads data onto the MIC, and the MIC processes these data, then the result is returned after completion. In the case of the CPU, the process starts the OpenMP processes according to the number of CPU cores to process these data. In each of the iterations, the MIC exchanges the boundary data with the CPU, and the adjacent processes need to communicate with each other; then the next iteration starts. After the iterations end, each of the processes sends results to the main process, which proceeds to output. In Fig. 11.29 we display the CPU+MIC hybrid process diagram of parallel LES on two nodes with two MICs in each node.

In each device (CPU or MIC), parallelism is implemented by OpenMP. The CPU and the MIC share the same core codes, and a different number of threads is transmitted to call the CPU or the MIC. Usually the number of threads on the CPU is the same as the number of CPU cores (when hyperthreading technology is turned off), or twice the number of CPU cores (when hyperthreading technology is turned on), while the number of threads on the MIC should be configured as three or four times more than the MIC cores.

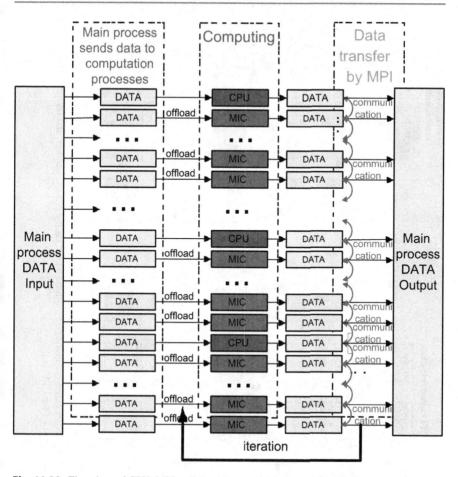

Fig. 11.28 Flowchart of CPU+MIC collaborative computing on multiple nodes

11.2.5.2 Data Division

In order to make it convenient to transfer the data among the processes after each iteration, we divide the LES data by lines (Fig. 11.30), and the solid line denotes the task division mode. According to the D2Q9 model, each grid point needs the data both above and below it to update itself. So, we need to allocate two lines of data for each process, which is displayed in the dotted line on the figure.

In this case, we adopt the static partitioning method to the case of parallel LES based on CPU+MIC on multiple nodes. First, according to the number of the nodes, we partition the mesh, using the method shown in Fig. 11.31. Then, assume the size of the mesh is (nx, ny), and the height of the grid in each node is:

$$H[0] = H[1] = \ldots = H[N\text{-}2] = [ny/N]$$
$$H[N\text{-}1] = ny\text{-}H[0] * (N\text{-}1)$$

$$(11.10)$$

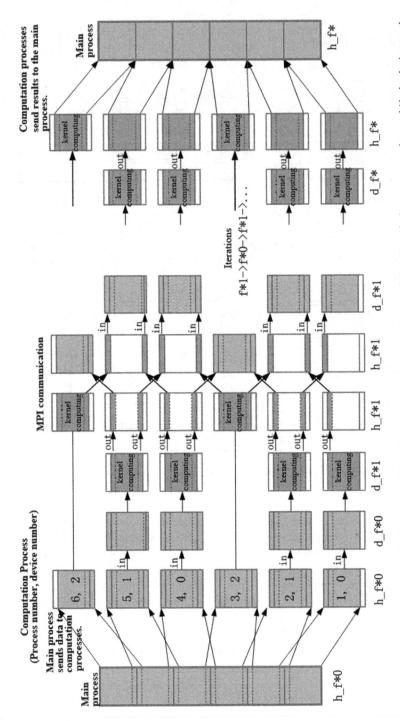

Fig. 11.29 Flow diagram of parallel LES based on CPU+MIC collaborative computing on multiple nodes. *i* is the process number, and *j* is the device number. The device number of the MIC is 0 and 1, and the device number of the CPU is 2

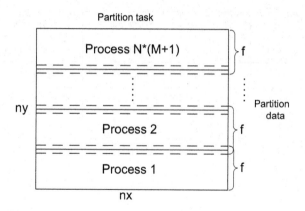

Fig. 11.30 Task division

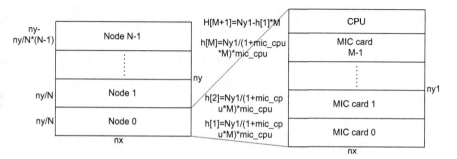

Fig. 11.31 Mesh division method

Because the computation capabilities of the CPU and the MIC are different, the computation complexity on the CPU and the MIC on each node cannot be divided equally. The height of the mesh on the MIC and the CPU is given, respectively, by:

$$h[1] = h[2] = \cdots = h[M] = \left[(int)\left(\frac{(float)(H[0])}{1 + mic_{cpu} * M} * mic_{cpu} \right) \right]$$

$$h[M + 1] = H[0] - h[1] * M$$

$$h[M + 2] = h[M + 3] = \cdots$$

$$= \left[h[(M + 1) * 2 - 1] = (int)\left(\frac{(float)(H[1])}{1 + mic\_cpu * M} * mic\_cpu \right) \right]$$

$$h[(M + 1) * 2] = H[1] - h[M + 2] * M$$

$$\cdots$$

Node ID	Device ID	Process number	Size of grid	Size of data
	M(CPU)	(M+1)*N	nx*h[(M+1)*N]	nx*(h[(M+1)*N]+2)
	M-1(MIC)	(M+1)*N -1	nx*h[(M+1)*N -1]	nx*(h[(M+1)*N -1]+2)
N-1	...	...	...	...
	1(MIC)	(M+1)*(N-1)+2	nx*h[(M+1)*(N-1)+ 2]	nx*(h[(M+1)*(N-1)+ 2]+2)
	0(MIC)	(M+1)*(N-1)+1	nx*h[(M+1)*(N-1)+1]	nx*(h[(M+1)*(N-1)+1]+2)
...	...	...	...	...
...	...	...	...	...
	M(CPU)	(M+1)*2	nx*h[(M+1)*2]	nx*(h[(M+1)*2]+2)
	M-1(MIC)	(M+1)+M	nx*h[(M+1) +M]	nx*(h[(M+1) +M]+2)
1	...	...	...	...
	1(MIC)	(M+1) +2	nx*h[(M+1) +2]	nx*(h[(M+1) +2]+2)
	0(MIC)	(M+1) +1	nx*h[(M+1) +1]	nx*(h[(M+1) +1]+2)
	M(CPU)	M+1	nx*h[M+1]	nx*(h[M+1]+2)
	M-1(MIC)	M	nx*h[M]	nx*(h[M]+2)
0	...	...	...	...
	1(MIC)	2	nx*h[2]	nx*(h[2]+2)
	0(MIC)	1	nx*h[1]	nx*(h[1]+2)

Fig. 11.32 The size of the data and computation complexity of CPU or MIC

$$h\big[(M+1) * (N-1) + 1\big] = h\big[(M+1) * (N-1) + 2\big] = \cdots$$
$$= h\big[(M+1) * (N-1) + M\big]$$
$$= \left[(int) \left(\frac{(float)(H[N-1])}{1 + mic\_cpu * M} * mic\_cpu \right) \right]$$

$$h[(M+1) * N] = H[N-1] - h[(M+1) * (N-1) + 1] * M \qquad (11.11)$$

Where mic_cpu denotes the ratio of the computation capability of one MIC to all the CPUs on one node. We introduce the mic_cpu in detail in the next section on load-balancing.

The computational complexity and amount of data controlled by each device (CPU or MIC) is shown in Fig. 11.32. The size of the memory allocated for the distribution function should be two lines larger than the grid. For the convenience of programming, the uppermost and the last processes should be allocated two lines memory more. One line of this will not be used, and the upper and lower boundary will be processed separately.

11.2.5.3 Load Balancing

Parallel LES on the CPU+MIC needs three levels of load balancing: load balancing among nodes, load balancing between the CPU and the MIC in one node, and load balancing among different threads on one device.

In the LES algorithm, because of the large number of iterations, if we partition the data dynamically, there will be plenty of communications between the CPU and the MIC in each iteration, which is not good for performance optimization. We therefore must partition data and achieve load balancing statically.

Load balancing among nodes

Because the computation load in every grid point is almost the same, the data can be partitioned equally in these grids. While the computation capability of each node is identical, the load balancing among nodes can be achieved. Every node controls nx*(ny/N) of the grids.

Load balancing between CPU and MIC in one node

Because there are some differences between CPU and MIC computation capabilities, the size of the grids allocated to the CPU must be different from those allocated to the MIC. We use mic_cpu to denote the ratio of one MIC to one multi-core CPU computation capability. We can draw the value of mic_cpu by the machine learning method. This means that if a certain size of grid is processed on both the CPU and the MIC, we can determine their running time. Then we obtain the value of mic_cpu by dividing the time on the CPU by the time on the MIC. Afterwards, we can partition the data in the grid according to mic_cpu value, which is shown in the right section of Fig. 11.31.

Load Balancing among different thread in one device

On the CPU or the MIC, we can achieve parallelism with OpenMP, and the computation complexity of each grid point is almost equal. We can also achieve load balancing statically among OpenMP threads. This means there is enough the default management of OpenMP.

11.2.5.4 Communication Optimization

There are only a few input variables in the LES algorithm, and not much reading from the hard drive. Hence the I/O operations need not be optimized.

In the parallel LES based on CPU+MIC, there are two aspects of the communication:

- Data transfer between adjacent processes
- Data transfer between the CPU and the MIC in the same process

Data transfer between adjacent processes

In the LES algorithm, the lattice points are divided into different processes according to where they are located. When each point updates itself and

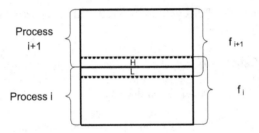

Fig. 11.33 Data transfer between adjacent processes

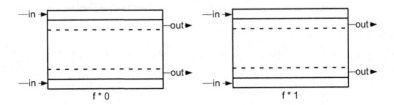

Fig. 11.34 Data transfer between MIC and CPU

migrates, the distribution function of the boundary lattice point in different fields should be transmitted to the adjacent processes. As shown in Fig. 11.33, the solid line between the dotted line H and L divides the computation field in two. But for the convenience of the migration computation of this solid line, the computation field of process i expands to the L grid boundary below, and the computation field of process i+1 expands to the H grid boundary above. The distribution function of H in the process i+1 should be transmitted to process i, and the process i+1 need to receive the distribution function of L from process i. The data transfer between adjacent processes is shown in the "MPI Communication" section in Fig. 11.29.

Data transfer between the CPU and the MIC in the same process

Because there is data transfer between the CPU and the MIC, before the first iteration the initial data to be processed on the MIC should be transferred from the CPU to the MIC. There is no such data transfer in the following iterations, and we only need to exchange the adjacent two lines (Fig. 11.34). The data transfer between the CPU and the MIC in the same process is shown in the "in" and "out" sections in Fig. 11.29.

Reduce the data transfer

After each loop is completed, data transfer must be processed between the CPU and the MIC in the same process and between processes. There are distribution functions in nine directions in every grid. In the kennel computation, we do not need all the values from these nine directions, and three directions suffice (Fig. 11.35). For process i, it only needs to receive the value of fsw, fs, and fse. It is the same for process i+1, which only need to receive the values of fnw, fn, and fne.

11.2.5.5 Pseudo-code for Parallel LES Based on CPU+MIC Collaborative Computing

The pseudo-code for the discussion in the previous sections is shown below:

```
[pseudo-code]
1    void main()
2    {
3        init();
4        Initialize MPI, start multiple MPI processes, and allocate arrays on each node.
5        if(myrank==root) //myrank denotes process number, and root denotes the main process.
6        {
7            Then main process send initialized data to every computation process;
8        }
9        else
10       {
11           The computation processes receive the initialized data from the main process;
12       }
13
14       if(myrank!=root) //the iterations of computation processes
15       {
16           deviceID = (myrank-1)%(M+1); //M denotes the number of MICs on one node.
17           /* Allocate space on MIC, and transfer the input data without releasing the space, ny denotes the
     height of grid in computation processes.*/
18   #pragma offload target(mic) \
19           in(fr0, fn0, ..., fse0: length(nx*ny) alloc_if(1) free_if(0)) \
20           nocopy(fr1, fn1, ..., fse1: length(nx*ny) alloc_if(1) free_if(0))
21       {}
22       for(i=0; i<ITR; i++)
23       {
24           if(i%2==0)
25           {
26               if(deviceID<M) //MIC kernel
27               {
28   #pragma offload target(mic) \
29                   nocopy( fr0, fw0, fe0, fr1, fw1, fe1) \
30                   in(fn0[0:nx],  fnw0[0:nx],  fne0[0:nx],  fs0[(ny+1)*nx:nx],  fsw0[(ny+1)*nx:nx],
     fse0[(ny+1)*nx:nx]: alloc_if(0) free_if(0)) \
31                   out(fs1[nx:nx],  fsw1[nx:nx],  fse1[nx:nx],  fn1[ny*nx:nx],  fnw1[ny*nx:nx],
     fne1[ny*nx:nx], fs1[(ny+1)*nx:nx] : alloc_if(0) free_if(0))
32               {
33                   /* THREAD_NUM_MIC denotes the number of OpenMP threads on MIC.*/
34                   LBCollProp(THREAD_NUM_MIC, ...); //migration and collision
35                   LBBC_LR(THREAD_NUM_MIC, ...); //left-right boundary
36                   if(myrank==totalranks-1) // totalranks denotes the number of all the processes.
37                       LBBC_UP(THREAD_NUM_MIC, ...); //upper boundary
38                   if(myrank==1)
39                       LBBC_DOWN(THREAD_NUM_MIC, ...); //lower boundary
40               }
```

```
41                  }
42                  else //CPU kernel
43                  {
44                      /* THREAD_NUM_OMP denotes the number of OpenMP threads on CPU*/
45                      LBCollProp(THREAD_NUM_OMP, ...); // migration and collision
46                      LBBC_LR(THREAD_NUM_OMP, ...); // left-right boundary
47                      if(myrank==totalranks-1) // totalranks denotes the number of all the processes.
48                          LBBC_UP(THREAD_NUM_OMP, ...); // upper boundary
49                      if(myrank==1)
50                          LBBC_DOWN(THREAD_NUM_OMP, ...); // lower boundary
51                  }
52                  /*MPI communication*/
53                  Communicate with the adjacent processes and interchange the boundary data;
54              }
55              else
56              {
57                  The same as the if branch, interchange the I/O data;
58              }
59          } //iterations end
60          if(deviceID<M) //Return the results from MIC and release the allocated space.
61          {
62              if(ITR%2==0)
63              {
64 #pragma offload target(mic) \
65                  out(fr0, fn0, ..., fse0: length(nx*ny) alloc_if(0) free_if(1)) \
66                  nocopy(fr1, fn1, ..., fse1: length(nx*ny) alloc_if(0) free_if(1))
67                  {}
68              }
69              else
70              {
71 #pragma offload target(mic) \
72                  nocopy(fr0, fn0, ..., fse0: length(nx*ny) alloc_if(0) free_if(1)) \
73                  out(fr1, fn1, ..., fse1: length(nx*ny) alloc_if(0) free_if(1))
74                  {}
75              }
76          }
77      }
78      if(myrank==root)
79      {
80          Receive the results from each computation process;
81          computeMacroparam (); //macroscopic parameters computation module
82          output(); //results
83      }
84      else
```

```
85        {
86                Send results to the main process;
87        }
88    }
```

Process i+1

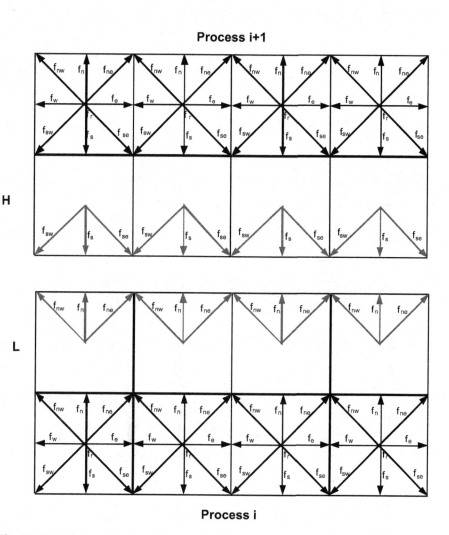

Process i

Fig. 11.35 Boundary migration of data transfer

11.2.5.6 Experimental Results

We show in Fig. 11.36 the experimental result of the parallel LES based on multiple nodes with the CPU+MIC. On a computing platform with two nodes, dual CPUs, and two MICs installed in each node, the performance of parallel LES is

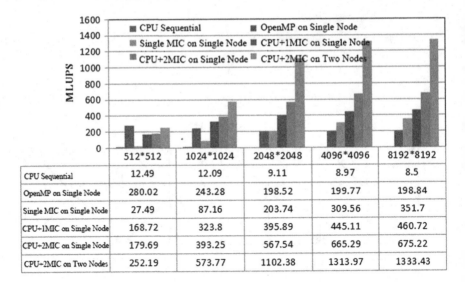

	512*512	1024*1024	2048*2048	4096*4096	8192*8192
CPU Sequential	12.49	12.09	9.11	8.97	8.5
OpenMP on Single Node	280.02	243.28	198.52	199.77	198.84
Single MIC on Single Node	27.49	87.16	203.74	309.56	351.7
CPU+1MIC on Single Node	168.72	323.8	395.89	445.11	460.72
CPU+2MIC on Single Node	179.69	393.25	567.54	665.29	675.22
CPU+2MIC on Two Nodes	252.19	573.77	1102.38	1313.97	1333.43

Fig. 11.36 Experimental results on multiple nodes with CPU+MIC

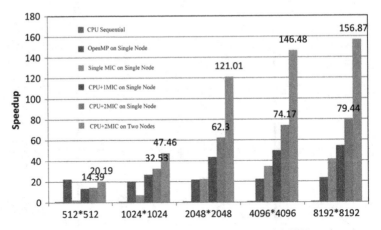

Fig. 11.37 Speedup of parallel LES on multiple nodes to sequential CPU version

1333.43MLUPS (when the mesh is 8192*8192 points). The speedup of the parallel LES on multiple nodes to sequential CPU version is shown in Fig. 11.37, and the maximum speedup in two nodes is 156.87. The speedup of parallel LES on multiple nodes to the OpenMP multi-thread version on a single node is displayed in Fig. 11.38, where we can see that the speed of a single MIC is 1.7 times faster than that of dual CPUs, and the speed of dual CPUs+2MICs is 3.4 times faster than that of dual CPUs, which means we have obtained an enhanced performance of 2.4

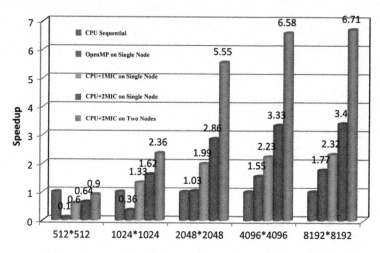

Fig. 11.38 Speedup of parallel LES on multiple nodes to OpenMP multi-threaded version

times improvement when two MICs are installed on a dual-CPU node. We can say that the performance of two nodes (dual CPUs+2MICs per node) is 6.71 times faster than the dual CPUs on a single node.

Further Reading

[1] Grama, M.: Introduction to Parallel Computing, 2nd edn. China Machine Press, Beijing (2005)

[2] Zhang, S., Yanli, C.: High Performance Computing on GPUs Using NVIDIA CUDA. China Water Power Press, Beijing (2009)

[3] Akhter, S., Roberts, J.: Multi-core Programming. Publishing House of Electronics Industry, Beijing (2007)

[4] Asia-Pacific Research and Development Ltd. Release the Potential of Multi-core. Tsinghua Press, Beijing (2010)

[5] Chen, G.: Design and Analysis of Parallel Algorithm. Higher Education Press, Beijing (2002)

[6] Kirk, D.: Programming Massively Parallel Processors. Tsinghua Press, Beijing (2010)

[7] Fu, X.: Improving program cache performance by program analysis and optimization. PhD thesis, (2007)

[8] Zhang, K., Zhang, Y., Hu, Z.: Development and frontier of electron microscopy 3D reconstruction. Acta Biophys. Sin. **26**(7), 533–559 (2010)

[9] Benzi, R., Succi, S., Vergassola, M.: The lattice Boltzmann equation: theory and applications. Phys. Rep. **222**(3), 145–197 (1992)

[10] under the armor of knights corner intel mic architecture/GeorgeChrysos. http://www.slideshare.net/IntelXeon/under-the-armor-of-knights-corner-intel-mic-architecture-at-hotchips-2012

[11] Intel Xeon Phi Coprocessor (Codename: Knights Corner) Software Developers Guide Ver. 1.04

[12] Intel C++ Compiler XE 13.0 User and Reference Guides

[13] Intel Fortran Compiler XE 13.0 User and Reference Guides

[14] Intel MIC Software Architecture KNC Technical Bootcamp

[15] Intel Xeon Phi Coprocessor Software Setup Guide

[16] Intel Xeon Phi Coprocessor OEM Software Configuration Tools Guide

[17] KNC ISA Reference Manual 1.0

E. Wang et al., *High-Performance Computing on the Intel® Xeon Phi™*,
DOI 10.1007/978-3-319-06486-4, © Springer International Publishing Switzerland 2014

Appendix: Installation and Environment Configuration of MIC

In the previous chapters, we introduced the software and hardware architectures of MIC. We now show the installation and configuration methods for the MIC environment, and readers should be ready to start compiling and running MIC programs by the end of this chapter.

Environment Configuration of MIC

Early Preparation

1. Hardware: Before the MIC environment installation, please check to make sure the hardware supports the MIC card and MIC environment.
2. Software: Software packages needed before installation:
   (a) Red Hat Enterprise Linux 6.0 GA 64-bit kernel 2.6.32-71
   (b) MPSS 2.1 Linux* Driver
   (c) l_ccompxe_intel64_2013.0.079.tgz
      The first package is the Linux system. The second is the MIC card driver, and the version above is the most recent one at the time this book was published (i.e., the version readers have may be different). The third package is the C/C++ compiler supported by MIC, which needs to be purchased from the agency, and the compiler version you purchase may be different. The newest version should be available on Intel's website: http://www.intel.com
3. BIOS Configuration: Enter BIOS, select "Advanced" > "Processor Configuration" > "Enhanced Intel SpeedStep Technology" > "Disable". This could improve performance.
4. RSA Key Generation: The RSA key needs to be set up before the MPSS is installed, not only for root authority, but also for user authority. This should ensure that the **mpssd** service obtains the keys automatically and copies them to the coprocessors when the service gets going.
5. IP Allocation: To ensure normal communication proceeds correctly, not only among the nodes in the host but also among the coprocessors outside the host, all the IP addresses allocated to the coprocessors should be configured in the same subnet. The same configuration should also be used in the other subnet of the cluster.

E. Wang et al., *High-Performance Computing on the Intel® Xeon Phi™*,
DOI 10.1007/978-3-319-06486-4, © Springer International Publishing Switzerland 2014

Installation of Linux on the Host

The Linux system on the client-side supplies users with the development and runtime environment for MIC programming. Users can select any of the operating systems below. Generally, users should choose one of the following:

- Red Hat* Enterprise Linux* 64-bit 6.0 kernel 2.6.32-71
- Red Hat* Enterprise Linux* 64-bit 6.1 kernel 2.6.32-131
- Red Hat* Enterprise Linux* 64-bit 6.2 kernel 2.6.32-320
- SuSE* Linux* Enterprise Server SLES 11 SP1 kernel 2.6.32.12-0.7-default
- SuSE* Linux* Enterprise Server SLES 11 SP2 kernel 3.0.13-0.27-default

Users can download the Linux operating system from an associated website or purchase it, but Intel does not supply any of these systems. You can also download the Linux operating system from one of the websites below:

http://www.suse.com/products/server/ or http://www.redhat.com/rhel/

Installation of the MIC Driver

The instillation of the Linux OS in not covered here. After the Linux OS is installed, we can install the MIC driver, which is available on the official Intel website. In addition, the MPSS software package version should be appropriate for the Linux version.

If the MPSS software package has been installed on the MIC and you would like to reinstall or update the package, the corresponding service should first be shut down and uninstalled. The detailed procedures are given below:

```
sudo service ofed-mic stop

sudo service opensmd stop

sudo service openibd stop
```

According to the hardware and driver, there might be no corresponding service.

```
sudo service mpss stop

sudo service mpss unload

sudo yum remove --noplugins --disablerepo=* intel-mic\*

sudo zypper remove intel-mic\*
```

Users can select the appropriate instructions. For example, you can use yum in Red Hat, and zipper in SuSE.

Now, the procedure for the driver installation:

1. Decompress KNC_beta_oem-2.1.3653-8-rhel-6.0.gz
     Instruction: tar KNC_beta_oem-2.1.3653-8-rhel-6.0.gz
2. Open the folder KNC_beta_oem-2.1.3653-8-rhel-6.0 and check the files included in this folder:

```
[root@mic4 beta2.1]# cd KNC_beta_oem-2.1.3653-8-rhel-6.0
[root@mic4 KNC_beta_oem-2.1.3653-8-rhel-6.0]# ls
gpl
intel-mic-2.1.3653-8.2.6.32-71.el6.x86_64.rpm
intel-mic-flash-0372-8.2.6.32-71.el6.x86_64.rpm
intel-mic-gpl-2.1.3653-8.el6.x86_64.rpm
intel-mic-knc-kmod-2.1.3653-8.2.6.32-71.el6.x86_64.rpm
intel-mic-sysmgmt-oem-2.1.3653-8.2.6.32-71.el6.x86_64.rpm
intel-mic-sysmgmt-oem-devel-2.1.3653-8.2.6.32-71.el6.x86_64.rpm
ofed
src
tic
```

3. Install the *.rpm files in this folder:

```
[root@mic4 KNC_beta_oem-2.1.3653-8-rhel-6.0]# sudo yum install --nogpgcheck --no
plugins --disablerepo=* *.rpm
```

When the following option appears, select "y" to begin installation:

```
Is this ok [y/N]: y
```

When the following results appear, it means the installation is complete:

```
Installed:
  intel-mic.x86_64 0:2.1.3653-8.el6            intel-mic-flash.x86_64 0:0372-8.el6
  intel-mic-gpl.x86_64 0:2.1.3653-8.el6        intel-mic-knc-kmod.x86_64 0:2.1.3653-8.el6
  intel-mic-sysmgmt-oem.x86_64 0:2.1.3653-8.el6   intel-mic-sysmgmt-oem-devel.x86_64 0:2.1.3653-8.el6

Complete!
```

4. Install OFED (optional):

The OFED software package is required when using IB net.

Before installation, please download OFED1.5.4.1 (the current driver only supports this version, and users should refer to the corresponding readme document in the MIC driver when installing):

http://www.openfabrics.org/downloads/OFED/ofed-1.5.4/OFED-1.5.4.1.tgz

When the download is complete, decompress it and open the folder. The instructions are:

```
[root@mic4 beta2.1]# cd OFED-1.5.4.1
[root@mic4 OFED-1.5.4.1]# ls
BUILD_ID install.pl ofed.conf.save RPMS    uninstall.sh
docs      LICENSE    README.txt     SRPMS
```

Install the install.pl file in this folder:

```
[root@mic4 OFED-1.5.4.1]# sudo perl install.pl
```

This process is simple, and users should select the options according to the following cues:

o Select option 2(Install OFED Software)

o Select option 3(All packages)

o Select option 1(Implementation: OFA)

o Press Y, Y, N

Then follow the instructions below:

sudo rpm –e kernel-ib

sudo rpm –e kernel-ib-devel

After installation, execute the following instructions:

sudo yum install --nogpgcheck   --noplugins –disablerepo=* ofed/*.rpm

When the following results appear, the installation is completed.

```
Installed:
  intel-mic-ofed-card.x86_64 0:3653-8.e16
  intel-mic-ofed-ibpd.x86_64 0:3653-8.e16
  intel-mic-ofed-kmod.x86_64 0:3653-8.e16
  intel-mic-ofed-kmod-devel.x86_64 0:3653-8.e16
  intel-mic-ofed-libibscif.x86_64 0:3653-8.e16
  intel-mic-ofed-libibscif-devel.x86_64 0:3653-8.e16

Complete!
```

5. Refresh Flash:
   (a) Make sure Intel-mic-flash rpm has been installed.
   (b) Run: /opt/intel/mic/bin/micinfo I grep –I "board stepping".
   (c) Confirm the status of the MPSS service by following the instructions below:

service mpss status

At this time ensure the service is closed.

   (d) Then refresh Flash:
       sudo /opt/intel/mic/bin/micflash –Update \
       /opt/intel/mic/flash/EXT_HP2_B0_0372-02.rom.smc
       B0 is the stepping sign of the MIC card. Please modify it according to
       your own MIC version. Before refreshing, use "sudomicctrl -w" to check if
       the card status is "ready". If not, follow the instructions below:

sudomicctrl –r

sudomicctrl –w

After refreshing Flash, restart the server.

6. Configuring the Intel® Xeon Phi™:
   (a) Initialize the configuration:

```
micctrl –initdefaults
```

    Then three files—default.conf, mic0.conf, mic1.conf—will be generated in /etc/sysconfig/mic/

   (b) Network configuration of MIC:
   • Modify default.conf:
     Remove the remark "# BridgeName micbr0".
     Set the bridge name to br0 on the host side.
   • Set the static IP address of every card:
     (i) Set the subnet according to the IP address of the host side, and allow the driver to set the subnet to 192.168. The IP address of the coprocessor should begin with 192.168.0.2:

```
# Define the first 2 quads of the network address.
# Static pair configurations will fill in the second 2 quads by default. The individual MIC
# configuration files can override the defaults with MicIPaddress and HostIPaddress.
Subnet 192.168.0.2


# Source for base of embedded Linux file system
BaseDir  /opt/intel/mic/filesystem/base  /opt/intel/mic/filesystem/base.filelist
```

    (ii) Add "MicIPaddress<your IP here>" in mic*.conf
       Node1 –mic0:

```
# Include configuration common to all MIC cards
Include default.conf


# Hostname to assign to MIC card
Hostname "node1 –mic0)"


# MIC MAC Address allocated and assigned by mpssd
MicMacAddress "ca:18:12:e5:45:70"
MicIPaddress   192.168.0.2
```

Node1 –mic1:

```
# Include configuration common to all MIC cards
Include default.conf

# Hostname to assign to MIC card
Hostname "node1 –mic1"


# MIC MAC Address allocated and assigned by mpssd
MicMacAddress "7a:20:73:57:ac:f7"
MicIPaddress  192.168.0.3
```

Note: The meaning of "node1 –mic0 192.168.0.2" is that the IP address of the first MIC card on node 1 is 192.168.0.2.

- Set the value of MTU:
  Add "MTUsize 9000" in default.conf.
- Set the netmask of every card:
  Add "NetBits 24" in default.conf.

(c) Mounted File System configuration:
- Check the file /opt/intel/mic/filesystem/base/etc/fstab, and ensure the mounted/home filesystem has been listed. For example:

```
192.168.0.254:/home /home nfs    rsize=8192,wsize=8192,nolock,intr    0 0
```

- Add "mount -a" to the end of /opt/intel/mic/filesystem/base/etc/rc.d/rc. sysinit.

(d) Client access on MIC:
  If the RSA key for the client has been established before the "service MPSS start" is called, the MPSS service will automatically set the client validation without password.

7. Start service:
  (a) Once all the configurations have been completed, the MPSS service can be started:

```
micctrl –resetconfig
sudo service mpss start
```

The result is:

```
[root@mic4 KNC_beta_oem-2.1.3653-8-rhel-6.0]# sudo service mpss start
Starting MPSS Stack:                                        [ OK ]
mic0: online (mode: linux image: /lib/firmware/mic/uos.img)


[root@mic4 KNC_beta_oem-2.1.3653-8-rhel-6.0]# service mpss status
mpss is running                          –
```

(b) Start the OFED-MIC service:

> Restart openied service
> Run "service ofed-mic start"

(c) Ensure the services are still running after restart
Note: If the computing node can't be found, check the service status.
Start the services automatically:

> chkconfig MPSS on : start MPSS service automatically
>
> chkconfig ofed-mic on : start OFED-MIC service automatically
>
> Till now, the driver has been installed successfully.

Installation of C/C++ Compiler on MIC

1. Decompress l_ccompxe_intel64_2013.0.079.tgz:
   tarzxvf l_ccompxe_intel64_2013.0.079.tgz
   Then, open the folder l_ccompxe_intel64_2013.0.079, and check the files in the folder using the ls instruction:

   ```
   [root@mic4 l_ccompxe_intel64_2013.0.079]# ls
   cd_eject.sh  install.sh  license  pset  rpms  support.txt
   ```

2. Modify some variables before running the install.sh file
   (a) Modify the "SELINUX=enforcing" to "SELINUX=disabled" in the path /
       etc/sysconfig/selinux. Then save and quit:

   ```
   [root@mic4 1 ccompxe intel64 2013.0.079]# vim /etc/sysconfig/selinux
   ```

   (b) Restart the server by the instruction: reboot.
3. Execute install.h in the directory l_ ccompxe_intel64_2013.0.079

   ```
   [root@mic4 1_ccompxe_intel64_2013.0.079]# ./install.sh
   ```

   Next, install the compiler. Users should select the appropriate installation option.
   When the third procedure starts, users must select the activation method. Here, we have chosen the second method, which selects the current license file.

> Please type a selection or press "Enter" to accept default choice [1]:2
> Note: Press "Enter" key to back to the previous menu.

   Input the whole license path to activate (according to the user's actual path).

Please type the full path to your license file(s):/root/intel/MIC_allkeys_Beta_XE2013.lic

----

Activation completed successfully.

Press "Enter" key to continue:

After activating the license, follow the instructions for installation accordingly. When the following results appear, the compiler installation is complete.

Please type a selection or press "Enter" to accept_default choice[q]:

Next, the environment variables should be configured after MPSS is installed: Use the instruction "more ~/.bashrc" to display the contents of .bashrc:

```
# .bashrc

# User specific aliases and functions

alias rm='rm -i'
alias cp='cp -i'
alias mv='mv -i'

# Source global definitions
if[ -f /etc/bashrc ]; then
        ./etc/bashrc
fi
```

Open the file .bashrc by vim, and add the following instructions. Then save and quit:

Source /opt/intel/composerxe/bin/compilervars.sh intel64

Update the environment variables:

[root@mic4 l_ccompxe_intel64_2013.0.079]# source ~/.bashrc

SDK Samples

The MIC environment has now been set up. To check how well the environment works, we can perform a test run: computing PI on the MIC card. If this program executes successfully, "PASS Sample01" appears. If there is something wrong, "*** FAIL Sample01" appears.

```
#include <stdio.h>
    #include <math.h>
    int main()
    {    float pi = 0.0f;
         int count = 10000;
         int I;
#pragma offload target (mic)
         for(i=0; i<count; i++)
         {
                 float t = (float)((i+0.5f)/count);
                     pi += 4.0f/(1.0f+t*t);
         }
         pi /= count;
         if(fabs(pi-3.14f)<=0.01f)
#ifdefDEBUG
                 printf("PASS Sample01 Pi = %f\n", pi);
         else
                 printf("*** FAIL Sample01 Pi = %f\n", pi);
#else
                 printf("PASS Sample01 Pi = %f\n");
         else
                 printf("*** FAIL Sample01\n");
#endif
    }
```

First, compile the program:

```
[root@mic4 wangyj]# icc -o PI PI.c -DDEBUG
```

Then execute it; we should get the following result:

```
[root@mic4 wangyj]# ./PI
PASS Sample01 Pi = 3.141593
```

Thus, we can determine whether the MIC environment has been set up successfully.

Index

Printed in the United States
By Bookmasters